OXFORD CHEMISTRY MASTERS

Series Editors

RICHARD G. COMPTON
University of Oxford

STEPHEN G. DAVIES
University of Oxford

JOHN EVANS
University of Southampton

OXFORD CHEMISTRY MASTERS

Circular Dichroism and Linear Dichroism

ALISON RODGER
University of Warwick

and

BENGT NORDÉN
Chalmers University of Technology, Gothenburg

Oxford   New York   Tokyo
OXFORD UNIVERSITY PRESS
1997

Oxford University Press, Great Clarendon Street, Oxford OX2 6DP
Oxford   New York
Athens   Auckland   Bangkok   Bogota   Bombay   Buenos Aires
Calcutta   Cape Town   Dar es Salaam   Delhi   Florence   Hong Kong
Istanbul   Karachi   Kuala Lumpur   Madras   Madrid   Melbourne
Mexico City   Nairobi   Paris   Singapore   Taipei   Tokyo   Toronto
and associated companies in
Berlin   Ibadan

Oxford is a trade mark of Oxford University Press

Published in the United States
by Oxford University Press Inc., New York

A catalogue record for this book is available from the British Library

Library of Congress Cataloging-in-Publication Data
Rodger, Alison.
Circular dichroism and linear dichroism / Alison Rodger and Bengt
Nordén.
(Oxford chemistry masters)
Includes bibliographical references.
1. Circular dichroism.   2. Linear dichroism.   I. Nordén, Bengt,
1945–   . II. Title. III. Series.
QD473.R6   1997   547.3'0856—dc20   96–43026
ISBN 0 19 855897 X (Hbk)

Typeset by the authors

Printed in Great Britain by
The Bath Press, Bath

Series Editors' Foreword

Oxford Chemistry Masters are designed to provide clear and concise accounts of important topics—both established and emergent—that may be encountered by chemistry students as they progress from the senior undergraduate stage through postgraduate study to leadership in research. These Masters assume little prior knowledge, other than the foundations provided by an undergraduate degree in chemistry, and lead the reader through to an appreciation of the state-of-the-art in the topic whilst providing an entrée to the original literature in the field.

In this first book of the Series, Alison Rodger and Bengt Nordén have produced an authoritative yet easy-to-read introduction to *Circular dichroism and linear dichroism*. This book will interest both novices, initiates, and masters in the field who wish to appreciate the importance of the powerful techniques described and to exploit them in the study of a wide diversity of chemical and biological phenomena.

Richard G. Compton
Stephen G. Davies
John Evans

Preface

The aim of this book is to provide an introduction for students and others who wish to use the techniques of optical activity (circular dichroism, *CD*) and optical anisotropy (linear dichroism, *LD*) for the study of the structure of molecules and interactions between molecules in solution. The emphasis is on what the techniques are and how to use them for both low and high molecular weight molecules. While there is a vast literature on *CD* with several excellent books produced over the years, the common request for a reference that will easily provide what an intending user needs to know is hard to answer. We hope that this text provides just such a guide. Further, we have aimed to take the use of linearly polarized light from being seen as rather exotic and only for experienced spectroscopists, to an approach to molecular structure elucidation that may be routinely and usefully implemented for a wide variety of problems.

The book has been constructed like a ramp: it begins with the basic principles and practices of *CD* and *LD* spectroscopy described in a way that does not require any advanced mathematics or quantum mechanics. The level of the subject matter gradually increases as the book proceeds, concluding with quantum mechanical derivations of the *CD* equations stated and used in the middle chapters. The central chapters should enable the reader to use, for example, exciton *CD* for identification of absolute configuration of simple bichromophoric molecules, *LD* for determining orientation of small molecules in anisotropic media and for assigning transition moment directions, and both techniques for investigating binding strengths and geometries of ligand–macromolecule complexes. Some of the advantages of combined *CD/LD* studies are also outlined.

We are extremely grateful for the interest, help, and tolerance shown by the people in our respective immediate surroundings. In particular this book would not have been finished without the encouragement of our families, the willingness of our students to read and comment on the text and to perform many of the experiments which now illustrate the text, and John Freeman who managed to convert rough sketches into artistic illustrations of chirality.

Warwick
Gothenburg
June 1996

A.R.
B.N.

To
Mark and Gunnela

Contents

6 Analysis of circular dichroism: magnetic dipole allowed transitions and magnetic *CD*

7 Circular dichroism formalism

Appendices

Index

1 Spectroscopy, chirality, and oriented systems

1.1 Introduction

Many of the questions we ask about chemical and biological systems relate to what shapes molecules have and how they interact with each other. New or modified methods are constantly being developed to address these questions, either in response to new problems or as the result of technical advances. The aim of this book is to show how circular dichroism (CD) and linear dichroism (LD) spectroscopies can be used to provide information about molecular structure and about interactions between molecules.

Although we have not yet defined what we mean by polarized light, it is still appropriate at the beginning of this book to define CD and LD. *Circular dichroism* is the difference in absorption, A, of left and right *circularly* polarized light:

$$CD = A_\ell - A_r \tag{1.1}$$

CD is particularly useful for studying chiral molecules, by which we mean ones that cannot be superposed on their mirror images.[1]

Linear dichroism is the difference in absorption of light *linearly* polarized parallel and perpendicular to an orientation axis:

$$LD = A_\parallel - A_\perp \tag{1.2}$$

Linear dichroism is used with systems that are either intrinsically oriented or are oriented during the experiment.

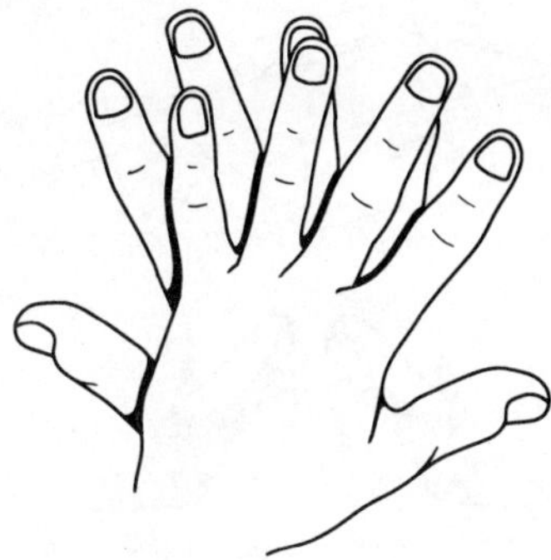

Chiral is derived from the Greek word χειρ meaning hand (hence the alternate term 'handedness'). Two molecules that are mirror images of each other are often referred to as *enantiomers*.

Fig. 1.1 Schematic illustration of *CD* and *LD*.

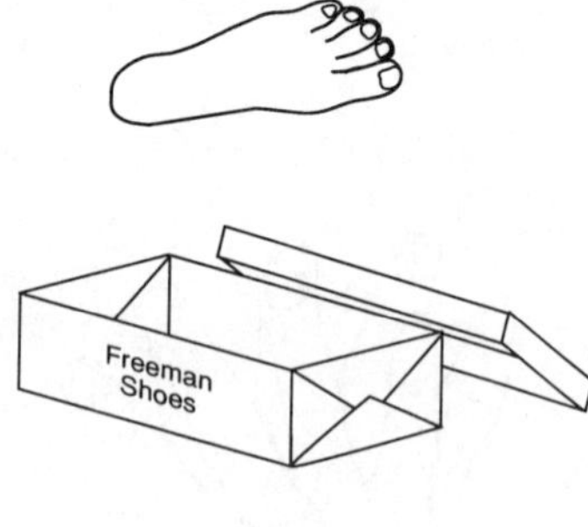

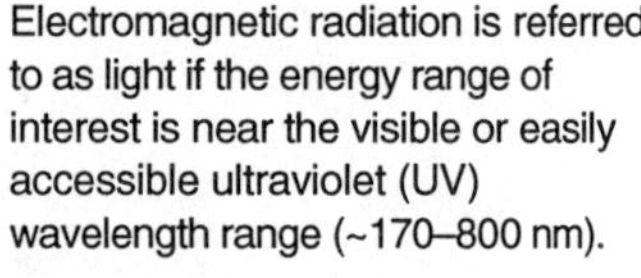

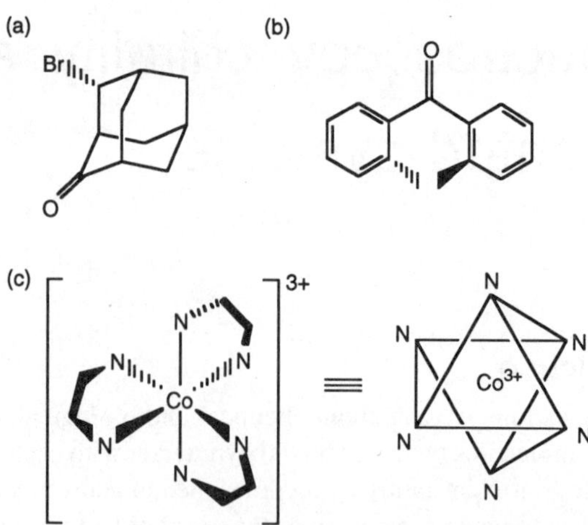

Fig. 1.2 (a) R-β-equatorial bromoadamantanone, (b) a chiral benzophenone conformation, and (c) Δ-[Co(en)$_3$]$^{3+}$, en = ethylenediamine.

The justification of our interest in *CD* and *LD* is that both chiral and oriented systems are intrinsic features of the world in which we live. All biological systems and many non-living ones are chiral; oriented molecular systems that lead to macroscopic effects include crystals, liquid crystals, and membranes. Logs flowing down a river and the muscles that control our bodies are larger scale examples of oriented systems. Sometimes the oriented system is itself built up by chiral molecules and the whole system may exhibit a macroscopic handedness.

Chapter 1 contains all that is necessary to understand what *CD* and *LD* are and to enable the techniques to be implemented in the laboratory. The emphasis of Chapters 2–6 is on the analysis of experimental data, and in the final chapter the theoretical foundation for *CD* is laid.

1.2 Electromagetic radiation and spectroscopy

Electromagnetic radiation

Electromagnetic radiation, as its name implies, has an electric and a magnetic field; these oscillate at right-angles to one another and to the propagation direction. Light may be described by a transverse wave whose *polarization* is defined by the direction of its electric field (Fig. 1.3). Electromagnetic radiation also has particle character. It is impossible to really understand this, we just have to accept the dual character of light. If we do an experiment that would probe wave-like behaviour, it is apparent that light has it, and if we look for particle character we find that too. Thus we refer to the *wavelength*, λ, and *frequency*, ν, of radiation, but acknowledge that if a molecule absorbs energy it absorbs discrete units called *photons*.

We have already referred to two types of polarized light: linear and circular. In a linearly polarized light beam all photons have their electric field, **E**, oscillating in the same plane, whereas in a circularly polarized light beam the electric field vector retains constant magnitude in time but traces out a helix about the propagation direction. Following the optics convention we take the end of the electric field vector of right circularly polarized light to form a right-handed helix in space at any instant of time. At each point in space or time, the magnetic field, **B**, is perpendicular to the electric field such that **k**, **E**, and **B** form a right-handed system as illustrated in Fig. 1.3.

If we view right circularly polarized light from a fixed point in space looking towards the light source, down the direction of propagation **k**, then the electric field vector traces out an anti-clockwise circle.

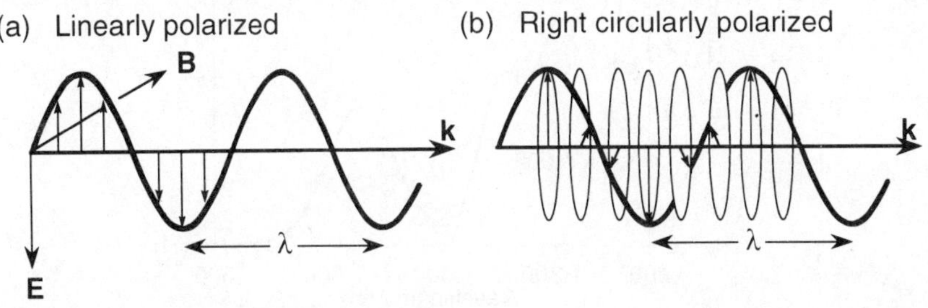

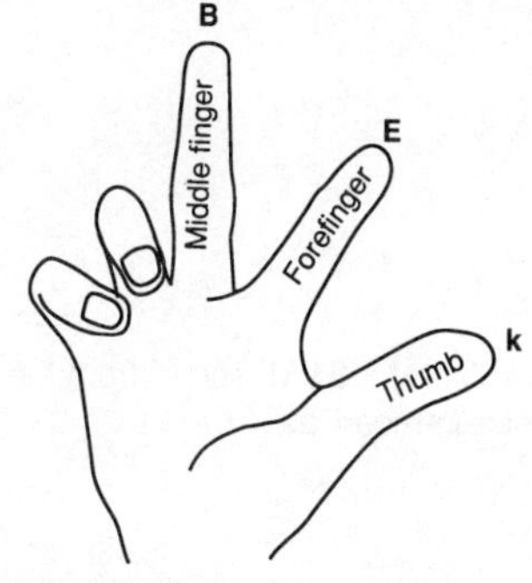

Fig. 1.3 (a) Linearly and (b) circularly polarized electromagnetic radiation. Arrows denote direction of **E**.

Normal absorption spectroscopy

Most spectroscopic phenomena arise from the interaction of a molecule with one photon at a time. The molecule either absorbs, emits, or scatters the photon. The simplest form of spectroscopy is absorption, where we measure how much light of a given frequency is absorbed by a collection of molecules. If a molecule absorbs a photon of frequency v, it increases its energy by

$$\Delta E = hv = \frac{h\omega}{2\pi} = \hbar\omega \qquad (1.3)$$

where h is Planck's constant and ω is known as the angular frequency. We shall concentrate on what happens when a molecule absorbs photons of visible or ultraviolet (UV) light since such a magnitude of energy [$(2\text{--}12)\times 10^{-19}$ J/molecule, $\lambda = 170\text{--}800$ nm] is required to rearrange the electron distribution in a molecule.

The Beer–Lambert law for the absorption, A, of light by a sample of concentration C is

$$A = \varepsilon C\ell \qquad (1.4)$$

where ℓ is the length of the sample through which the light passes, ε is known as the extinction coefficient; if ℓ is measured in cm and C in M $=$ mol dm^{-3}, then ε has units of mol^{-1} dm^3 cm^{-1}.

Absorbance is defined in terms of the intensity of incident, I_0, and transmitted, I, light:[2–4]

$A = \log_{10}(I_0/I)$.

The Beer–Lambert law is valid as long as the spectrometer can measure I and there are no concentration-dependent intermolecular interactions.

Absorption can be pictorially viewed as either the electric field or the magnetic field (or both) of the radiation pushing the electron density from a starting arrangement to a higher energy final one. The direction of net linear displacement of charge is known as the *polarization of the transition*. The polarization and intensity of a transition are characterized by the so-called *transition moment*, which is a vectorial property (see Appendix 1) having a well-defined direction (the transition polarization) within each molecule and a well-defined length (which is proportional to the square root of the absorbance). The transition moment may be regarded as an antenna by which the molecule absorbs light. Each transition thus has its own antenna and the maximum probability of absorbing light is obtained when the antenna and the electric field of the light are parallel.

ε(466 nm) = 84 M^{-1} cm^{-1} from the Beer–Lambert law.

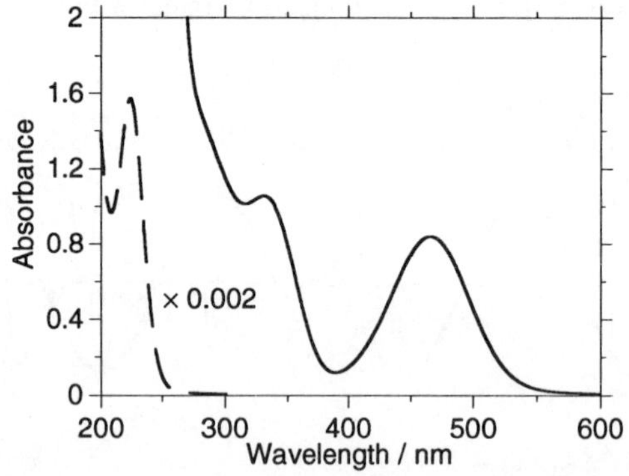

Fig. 1.4 Normal absorption spectrum of 0.1 M $[Co(en)_3]^{3+}$ (*cf.* Fig. 1.2) with ℓ = 0.1 cm.

The *bandwidth* of an electronic absorption band (usually defined as the width at half the maximum height) is not infinitely narrow but broadened due to:

- the uncertainty principle

$$\Delta E \Delta t \geq \frac{h}{4\pi} = \frac{\hbar}{2}$$

where ΔE is the uncertainty in the energy and Δt is the lifetime of the excited state
- Doppler broadening
- intermolecular interactions
- rotational transitions
- vibrational transitions.

In practice, in a collection of molecules, the photons absorbed by different molecules will be of slightly different energies so what we measure is a curve like the one in Fig. 1.4, where the signal that is plotted is a measure of the probability that a transition will occur at that energy. Such a plot of the absorbance of light versus λ or ν is known as an *absorption spectrum*.

Circular dichroism spectroscopy

Now consider chiral molecules—because a chiral molecule has no reflection plane, any rearrangement of its electrons will not have one either, so the electrons move in some kind of helix. From the fact that in circularly polarized light the electric field vectors trace out helices (Fig. 1.3) we realize that the interaction between a chiral molecule and left- and right-handed photons will be different. This is the idea behind the definition of *CD*.

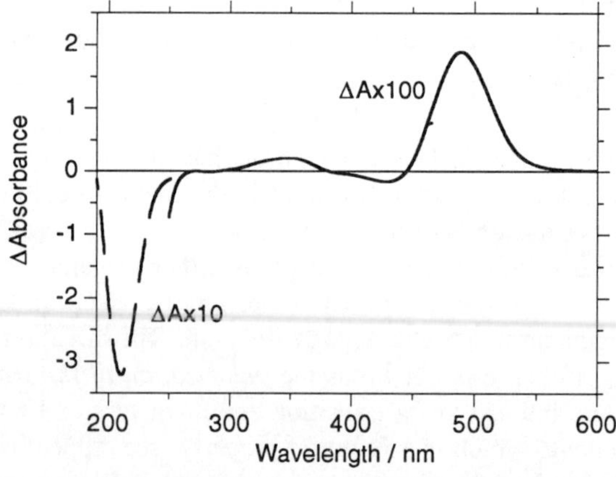

Fig. 1.5 *CD* spectrum of 0.1 M Λ-$[Co(en)_3]^{3+}$ in a 0.1 cm pathlength cell.

Even in an 'obviously' helical molecule different electronic transitions involve electron redistributions of different handedness so any one molecule will have both positive and negative *CD* signals (Fig. 1.5). The challenging task is then to relate the helical motions of the electrons to the arrangement of the atoms and bonds in space. This forms the subject matter of Chapters 2, 5, and 6.

CD is now a routine tool in many laboratories. The most common applications include proving that a chiral molecule has indeed been synthesized or resolved into pure enantiomers and probing the structure of biological macromolecules, in particular determining the α-helical content of proteins.

Linear dichroism spectroscopy

We are now in a position to understand *LD* a little better. Imagine we have a linearly polarized light beam (Fig. 1.3) and a sample of molecules all oriented in exactly the same way. If we first measure the normal absorption with the light polarized so that it is parallel to the direction of orientation of the sample, then measure it when the light is polarized perpendicular to this direction, the difference of these two spectra is an *LD* spectrum. What does it tell us? The two extreme cases are both illustrated in the simplified anthracene spectrum of Fig. 1.6.

It is tempting to imagine the helix traced out by **E** of a photon pushing the electrons along the helix of a transition. However, **E** is perpendicular to the helix axis *and* the wavelength of light is ~2000 times the size of a molecule.

The *LD* spectrum of Fig. 1.6 is simplified in two ways:
- perfect orientation of the molecules has been assumed, and
- the long wavelength band has been assumed to be of pure short-axis polarization.

In reality perfect orientation is never achieved and furthermore there is a significant long-axis polarized component at ~320 nm due to coupling with the 250 nm band.

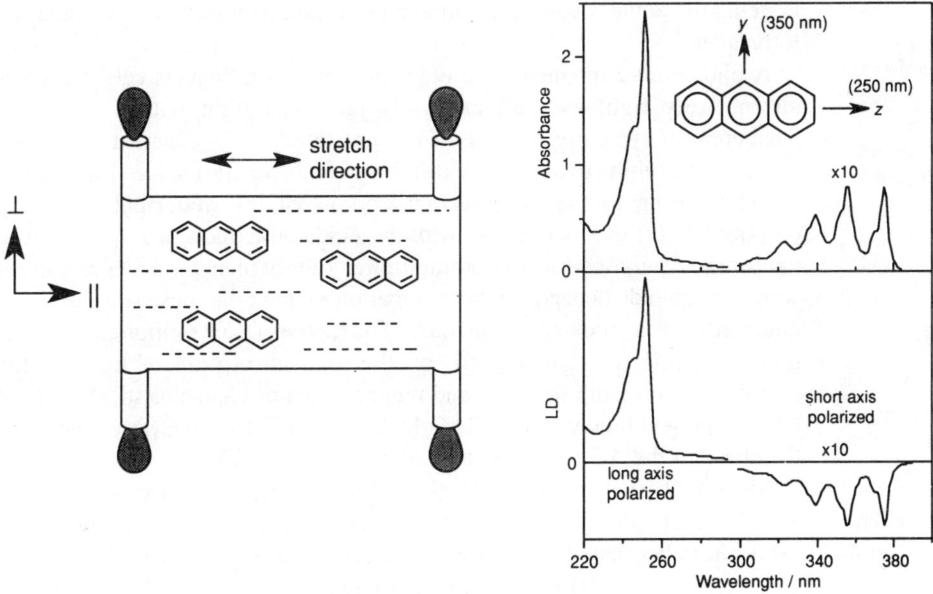

Fig. 1.6 Stretched film orientation and *LD* spectrum of long (*z*) and short (*y*) axis polarized transitions.

(1)   If the polarization (*i.e.* direction of the electric transition moment) of the transition we are probing is perfectly *parallel* to the orientation

direction, as would be the case for a transition polarized parallel with the long axis of a rod-shaped molecule when the orientation comes from putting the molecules in a film and stretching it (Fig. 1.6), then

$$LD = A_{\parallel} - A_{\perp} = A_{\parallel} > 0 \tag{1.5}$$

(2) If the polarization of the transition we are probing is *perpendicular* to the orientation direction, as would be the case for a pure short axis polarized transition of the molecule in the stretched film (Fig. 1.6), then

$$LD = A_{\parallel} - A_{\perp} = -A_{\perp} < 0 \tag{1.6}$$

For intermediate polarizations, the LD is between these cases. Thus, we can find out what the polarization is for a given transition from its LD spectrum if we know how the molecule is oriented; conversely, we can also use LD as a probe of molecular orientation if we know the polarization of a transition moment within the molecule. In practice, we often do one type of experiment and then the other using different orientation methods.

1.3 Measuring a *CD* spectrum

Circular dichrometers (spectropolarimeters)

The essential features of a CD spectropolarimeter are a source of (more-or-less) monochromatic left and right circularly polarized light and a means of detecting the difference in absorbance of the two polarizations of light. The normal method of achieving these requirements, because the CD of molecules is generally quite weak, is to implement a polarization phase-modulation technique.

A photoelastic modulator (in older instruments a Pockels cell) produces alternatively right and left circularly polarized light with a switching frequency of typically 50 kHz. The light intensity is constant, but upon passage through a sample exhibiting CD an intensity fluctuation (corresponding to the different absorptions of left and right circularly polarized light) that is in phase with the modulator frequency appears. The unabsorbed photons hit a photomultiplier tube which produces a current whose magnitude depends on the number of incident photons. This current is detected by a lock-in amplifier. Thus, the DC component of the photomultiplier current depends on the total absorption of light by the sample (and on lamp intensity and monochromator characteristics), and the AC component relates to the CD (a lock-in amplifier is needed to determine the phase of the AC component, *i.e.* the sign of the CD).

Most CD spectropolarimeters, although they measure differential absorbance, produce a CD spectrum in units of ellipticity, θ, in millidegrees, *versus* λ, rather than ΔA *versus* λ. The conversion between these two is:

$$
\begin{aligned}
CD = \Delta A(\text{absorbance units}) &= A_{\ell} - A_r \\
&= \frac{4\pi\theta(\text{degrees})}{180 \ln 10} \\
&= \frac{\theta(\text{millidegrees})}{32{,}982}
\end{aligned}
\tag{1.7}
$$

The voltage on the photomultiplier tube is larger when the number of incident (*i.e.* non-absorbed) photons is smaller, thus the high tension voltage on a *CD* machine is a *rough* guide to the absorbance of a sample.

'Ellipticity' dates from the time when the *CD* was quantitated in terms of the change of *polarization* of linearly polarized light into elliptically polarized light when passed through the sample (this was measured by a compensator technique). Today one measures instead changes of *intensity, i.e.* the true differential absorption of light. Therefore ellipticity units can be regarded as obsolete and should be replaced by molar absorbance units.

The *CD* version of the Beer–Lambert law [eqn (1.4)] is:

$$\Delta A = (\Delta \varepsilon) C \ell \qquad (1.8)$$

From Fig. 1.5 it follows that $\Delta\varepsilon$(489 nm) = +1.89 mol^{-1} dm^3 cm^{-1} for the left-handed screw form (Λ) of [Co(en)$_3$]$^{3+}$.

Some aids to collecting good *CD* data are as follows.

- Always run a normal absorption spectrum of both the sample *and* the solvent or buffer before collecting *CD* data since the concentration and wavelength ranges are the same for both types of spectrum. Do *not* collect the normal absorption spectra against references of the solvent/buffer since *CD* spectropolarimeters are single beamed. If the sample plus solvent/buffer has a large absorbance signal, say greater than 2 absorbance units, try using a shorter pathlength cell together with a more concentrated sample or use a different solvent/buffer.

- *All* of the light beam incident upon the cell must pass through the sample and not be clipped or reflected by the walls or base of the cell or the meniscus of the solution (otherwise the measured spectrum is affected by scattering). The narrow cells often used to minimize sample volume in normal absorption spectrophotometers *cannot* be used for *CD* unless the light beam is chopped by a mask or focused prior to hitting the cell.

- Either cylindrical or rectangular cuvettes are used for *CD*. Cylindrical cells usually have lower birefringence (baseline *CD*) and may require slightly less sample. Rectangular cells are cheaper, may be used in standard absorption spectrophotometers (so *CD* and normal absorption data may be collected on exactly the same sample), and may be used for serial titration experiments as ~60% of a rectangular cell can be empty for the first spectrum and gradually filled.

- The baseline is rarely straight in a *CD* experiment (it depends on the spectropolarimeter and the cuvette), so always collect a baseline spectrum of the solvent/buffer under the same conditions as the sample spectrum using the same cuvette in the same orientation with respect to the light beam. Subtract the baseline spectrum from the sample spectrum to produce the final *CD* plot.

- *CD* spectrometers usually scan from longer wavelengths to shorter ones. Select a wavelength range starting at least 20 nm beyond the normal absorption envelope. When the baseline spectrum is subtracted from the sample spectrum, the region outside the absorption envelope should be flat.

- Most *CD* spectropolarimeters have both short timescale (millisecond to minutes) and long timescale (minutes to hours) baseline variations. If the *CD* signal is large both can be ignored. To avoid any problem from short timescale variations collect data averaged over a number of faster scans rather than one slow one. Longer timescale fluctuations can be more problematic and are usually dealt with by alternating collection of sample and baseline spectra.

- It is usually the case that with small *CD* signals, even when the baseline is subtracted, the *CD* is not exactly zero outside the absorption

Titration experiments where spectra are collected as a function of concentration, ionic strength, pH *etc.* often involve adding solution to the cuvette. A simple way to avoid dilution effects is as follows. Consider a starting sample that has concentration x M of species X. Each time y μL of Y is added, also add y μL of a $2x$ M solution of X. The concentration of X remains constant at x M.

envelope. However, if it is flat it may be regarded as a linear base-line shift.

- Sample concentration should usually be such that the total absorbance of the sample is between ~0.2 and ~1.5 absorption units. Samples with a total absorbance above 2 often give an unreliable *CD* signal; an absorbance of 1.1 units is theoretically optimal. For high absorbances one should check that the *CD* signal is proportional to sample concentration by running a spectrum of a diluted sample.

- *CD* spectropolarimeters give the operator considerable control over the time constant, τ (time over which the machine averages data), scan speed, s, and bandwidth, b (the wavelength range of incident light). The signal to noise ratio increases as $\tau^{-1/2}$ so select τ as large as possible subject to $\tau \times s \leq \frac{b}{2}$. If τ is too long for the chosen s and b, then maxima of peaks (both positive and negative) will be cut off and their wavelengths shifted. A control scan using $\frac{\tau}{2}$ should be used to check that spectra are not being distorted by the chosen parameters.

- Signal to noise ratio increases roughly as the *square root* of: the number of scans, the time constant, and the intensity of the light beam.

- A fast preliminary scan will indicate whether there is any point in collecting an accurate spectrum and whether the sensitivity scale has been chosen correctly to appropriately display the *CD* at all wavelengths of the spectrum.

1.4 Design and implementation of an *LD* experiment

The components of an *LD* experiment are: a source of linearly polarized light, a means of detecting how much light is absorbed, a method of orienting the sample, and a way to change the relative orientations of sample and light beam.

In this book we are concentrating on measuring either the *CD* of *chiral* molecules in *isotropic* solutions or the *LD* of achiral (and also chiral) molecules in *anisotropic* solutions. It should be noted that while the intrinsic *CD* of a chiral molecule does not generally provide any problem for the measurement of *LD* in an oriented sample, the same is not true for measuring the *CD* of an anisotropic (*i.e.* oriented) sample. The reason for this is partly because the modulation technique discussed above for measuring *CD* spectra operates with continuously changing polarizations of light, and in addition to circular components there are linear ones present as well. Since the linear birefringence and *LD* of an anisotropic sample are usually one or two orders of magnitude larger than the *CD*, the latter is easily 'drowned' by the linear effects.

Measurement methods

There are two main methods for measuring *LD* spectra. The one requiring less specialized equipment is the two-spectra method. The differential method is much easier to implement and a wide range of sample orientation

Some instruments use response time, $\rho = 2\tau$. If $b = 2$ nm and $\lambda = 350$ nm, then the light ranges from 348 to 352 nm.

CD measurements on anisotropic samples require meticulous care including correction for sample imperfections as well as instrument optics.[5]

techniques may be used. However, much more sophisticated instrumentation is required with this method.

Two-spectra method for measuring LD

For measurement of strongly dichroic samples the simplest method of measuring *LD* is to use a double beam spectrophotometer equipped with a polarizer (*e.g.* of Glan type) which is not sensitive to beam divergence. The sample is oriented parallel to the polarization direction of the polarizer to obtain $A_{\parallel}$. It is then oriented perpendicular to the polarization direction of the polarizer to obtain $A_{\perp}$. (Alternatively, the polarizer may be rotated, in which case one must consider any effect of internal polarization of the light by the optics of the spectrometer.) It is advisable to put identical polarizers into both the sample and the reference beam of the spectrometer to improve the baseline and measurement sensitivity.

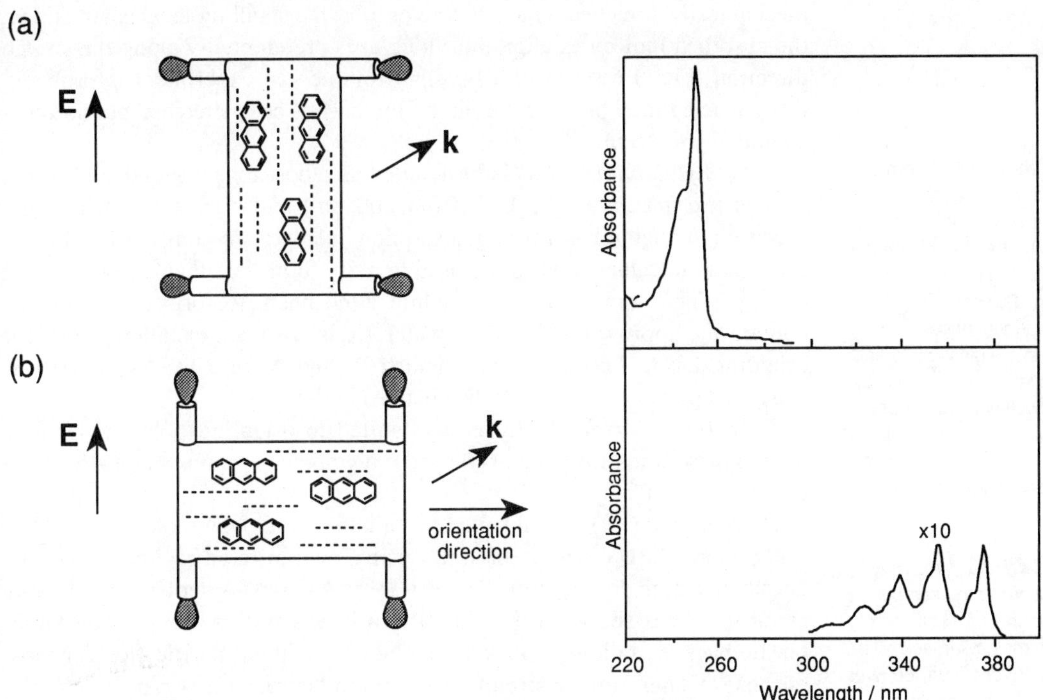

Fig. 1.7 Principle of the two-spectra method for measuring *LD*. (a) $A_{\parallel}$, (b) $A_{\perp}$ (see comment on Fig. 1.6).

Differential method for measuring LD

LD spectra may also be measured directly as a differential absorption using the phase-modulation technique of a circular dichrometer, either by supplementing the latter with a quarter wave device (*e.g.* a Fresnel rhomb or an Oxley prism to achieve achromacy[5]) or by increasing (by a factor of 2) the driving voltage of the photo-elastic modulator and doubling the beat frequency of the lock-in amplifier. The differential method has the advantage of being extremely sensitive, needing only very low sample orientation or concentration. However, for any quantitative analysis of the spectrum an independent measurement of the isotropic absorption, A, is required for the same sample. Since this measurement is generally performed using a different instrument (a dual beam absorption spectrometer) care has to be taken to avoid artefacts due to variations in wavelength, stray light, scattering, *etc.* all of which require careful calibration of the monochromators.

Molecular alignment techniques

The orientation method used depends on the sample. Long polymers may be oriented by shear flow whereas small molecules require a stronger orienting force. A selection of orientation methods is described below. More details may be found in references 6–8.

Stretched polymers

Small molecules can often be absorbed into polymer films. When the film is mechanically stretched, either before or after the small molecules are added, the absorbed molecules align their long axes preferentially along the stretch direction (Fig. 1.6). The best baseline for such an experiment is probably that collected for a piece of the same film that has been stretched but does not contain the molecule of interest.

Polyvinylalcohol (PVA) is best suited as a host for polar molecules; the film is transparent in the UV (from 200 nm) and visible regions of the spectrum, though it has a strong absorption over large regions in the infrared. For small molecules it has to be used in a dry form (less than a few per cent water). When equilibrated in a humid atmosphere to form an elastic gel containing approximately 50% water, it is also an excellent host for orienting DNA. *LD* spectra of oriented B- and A-form DNAs, as well as denatured polynucleotides may be obtained in this way.[8]

Polyvinylchloride (PVC) is also suited to orienting polar molecules, though these films are more brittle and more prone to rupture upon stretching than PVA and PE.

Polyethylene (PE) is well suited for orienting non-polar molecules and has transparency in UV (from 200 nm), visible, and infrared regions. Both low density and high density polyethylenes are useful, the former being preferable as most solutes show better solubility in it. Usual commercial transparent plastic bags or stationery pockets can be used without purification for most purposes. The film is stretched (at room temperature) parallel to the manufacturer's stretch direction by a factor of 4–8. It is advisable not to stretch too close to the breaking point of the polymer since the film has a tendency to become opaque. The solute may be introduced (doped) into the stretched film by soaking it in a saturated solution of the solute in cyclohexane or chloroform (or by adding a droplet of the solute in solution to the surface of the film then covering it with a microscope fused-silica cover slip). Chloroform also enters and swells the polyethylene and is itself aligned when the film is stretched, so the baseline must be measured on a film that has been treated in the same way as the sample film.

Note that there has been no evidence that a solute is better oriented when introduced into the film before stretching compared with it being added after

PVA films may be prepared from well-hydrolysed commercial PVA by dissolving the powder in initially cold water (10% w/v). The slurry is then heated to near boiling to form a viscous solution. The sample (~5 mM solution) is then added, and the mixture is cast onto a glass plate and allowed to dry over a period of one to three days in a well-ventilated dust-free place. Finally the film is stretched by a factor of 2–5 at an elevated (hot air from a hair dryer) temperature.

Polyvinylchloride (PVC) films may be prepared by dissolving the sample (0.1–2 mg) in 10 cm^3 of a solution of PVC (10 g, high molecular weight) in tetrahydrofuran (75 cm^3 distilled over $MgSO_4$ + $FeSO_4$). The resulting syrup is then transferred to a glass plate and the solvent allowed to evaporate at a controlled partial pressure (in a glass container with semi-closed lid) over three days. Finally the film is stretched by a factor of 2–5 at 60°C.

stretching. The orientation of the solute molecules, hence, appears to be caused by their adsorption under equilibrium conditions to aligned polymer chains or crystallites, or to the occupancy of anisotropic cavities.

Flow orientation

Long polymers may be oriented by the viscous drag caused when a solution is flowed between narrow walls. Depending on the cell design, the light is then propagated either along the flow direction or perpendicular to it. This technique is commonly used for *LD* studies of DNA (the DNA must be at least of the order of 1000 base pairs in length to get significant orientation).

The most successful flow cell has proved to be a cylindrical couette flow cell where the solution containing the DNA is subjected to a constant gradient over the annular gap between two coaxial cylinders one of which is rotating (Fig. 1.8). The speed of rotation must be large enough to cause significant orientation but not so large as to cause turbulent flow. Since the baseline may depend slightly on the position of the rotor, the best baseline to subtract from a recorded spectrum is that of the sample in the same cell but rotating too slowly to cause measurable orientation of the sample. Alternatively, a spectrum of the solvent/buffer with higher rotation speed may be measured as the baseline.

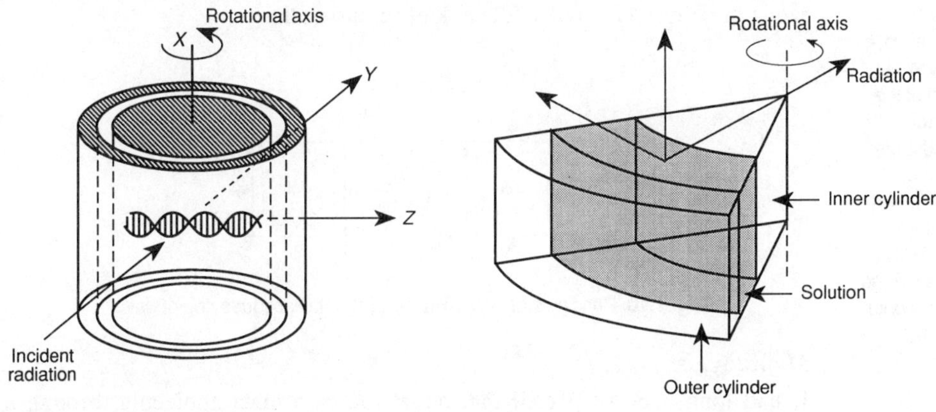

Fig. 1.8 Flow orientation in a coaxial (couette) flow cell with radial incident light. The total light path is typically 1 mm.

Electric field orientation

Effective orientation for polar or polarizable molecules may be achieved by an electric field between two parallel plates which produces uniaxial orientation of the molecules (Fig. 1.9). Electric orientation is conceptually simpler than flow orientation, and the data can be fairly easily extrapolated to infinite orientation which contrasts with the situation for flow orientation. Heating effects may be circumvented by using pulsed field techniques, which additionally enable the study of relaxation phenomena. The pulsed nature of the field, however, generally requires each electric dichroism experiment to be

DNA may be oriented by an electric field, although it is charged rather than dipolar, since its ionic environment is polarizable and establishes a net dipole in the presence of the field.

performed at a fixed wavelength or else a diode array system to be used; it is also usual to measure $A_{\parallel}$ and A (the unoriented absorbance) rather than $A_{\perp}$.

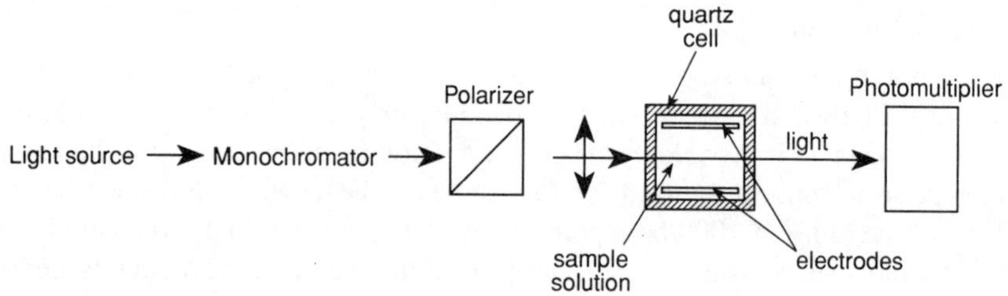

Fig. 1.9 Schematic illustration of an electric dichroism experiment.

Squeezed gel orientation

The method of orientation that has proved most successful for membrane proteins, particularly photosynthetic ones, is to embed the protein in a gel and then physically squeeze the gel either bidirectionally or unidirectionally resulting in respectively rectangular or square gels. The main shortcomings of the gel orientation technique are its limitation to wavelengths longer than 250 nm due to scattering and absorption by the gel, and its limited dynamic range of deformation due to the risk of rupturing the gel.

To make a polyacrylamide gel, to orient for example a photosynthetic pigment, a buffer suspension of the photosynthetic reaction centre particles is mixed with acrylamide (0–15% w/v); N,N'-methylene-bisacrylamide (0.3–0.5% w/v); and glycerol (50% v/v). The mixture is then polymerized by adding 0.03% (v/v) N,N,N',N'-tetramethylethylene-diamine and 0.05% (w/v) ammonium persulfate.[9]

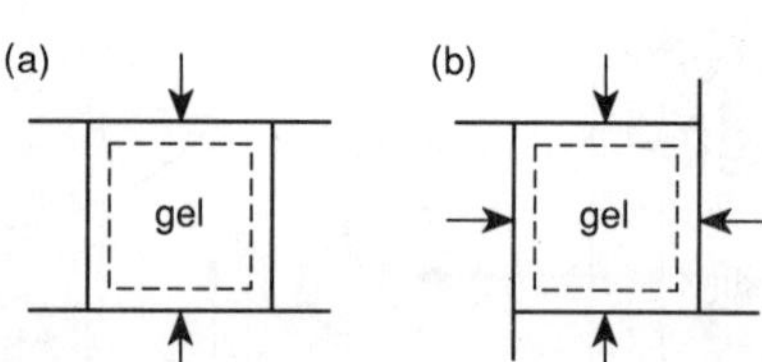

Fig. 1.10 Principles of (a) biaxial and (b) uniaxial squeezing of gels.[7]

Migrative orientation

It had long been suspected that migration of a macromolecule through a viscous or gel medium could create macroscopic orientation. It was not until 1985, however, that the electrophoretic orientation of DNA in a gel was demonstrated using LD.[10] Such studies have since given valuable information about the mechanisms underlying the success of pulsed fields to separate large 'reptating' DNA molecules in gel electrophoresis.[11]

Magnetic field orientation

Magnetic fields may be used to orient molecules in much the same way as electric fields, however, the effect is small and this method is not widely used unless the particles are large (such as chloroplasts) and carry substantial magnetic dipole moments.

Crystalline samples

If the absorbance is not too large for the spectrometer to measure, crystals may be used for *LD* studies. Alternatively, a polarized reflection spectrum of the thick crystal may be converted to the polarized absorption spectrum using a Kramers–Kronig transform.[12]

Liquid crystal orientation

The molecules within a liquid crystal are oriented with respect to one another and by sandwiching the liquid crystal between two quartz plates, macroscopic orientation may be achieved. Further, a molecule dissolved in the liquid crystal may also have a preferential orientation. This makes a technically very simple means of obtaining an oriented sample for *LD* spectroscopy.

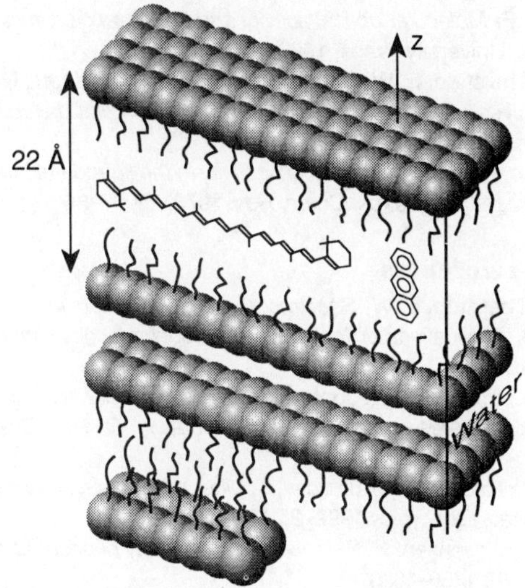

Fig. 1.11 Liquid crystal orientation for hydrophobic molecules. With the illustrated liquid crystal, molecules smaller than the bilayer thickness align parallel to the hydrocarbon chains.

The lamellar phase of the system sodium *n*-octanoate (18.2% w/w)/ *n*-decanol (35.5% w/w)/water (46.3% w/w) sandwiched between quartz plates forms just such a system with alternating layers as indicated in Fig. 1.11. In this case small hydrophobic molecules are oriented parallel to the long hydrocarbon chains just as in a polyethylene film. Ionic species may also become oriented, either parallel to the hydrocarbon chains, for an amphiphilic molecule, or perpendicular to the orientation axis but in the hydrophilic water layer.[13] Since there is no single direction of preferred orientation in the plane of the sandwiched quartz plates and the unique axis is perpendicular to the quartz plates, the *LD* measurement has to be performed at inclined incidence, *i.e.* by tilting the quartz plates relative to the light beam of the spectrometer.

Orientation by evaporation onto quartz

This method may be appropriate for some samples, particularly planar aromatic molecules that readily adsorb onto quartz. Just like the lamellar liquid crystal sample, the quartz plate has to be tilted for *LD* measurement.

References

General *CD* references

- Barron, L. D. *Molecular light scattering and optical activity.* Cambridge: Cambridge University Press, **1982**
- Craig, D. P.; Thirunamachandran, T. *Molecular quantum electrodynamics: An introduction to radiation-molecule interaction.* London: Academic Press, **1984**
- Harada, N.; Nakanishi, K. *Circular dichroic spectroscopy: exciton coupling in organic stereochemistry.* California: University Science Books, **1983**
- Mason, S. F. *Molecular optical activity and the chiral discrimination.* Cambridge: Cambridge University Press, **1982**
- Michl, J.; Thulstrup, E. W. *Spectroscopy with polarized light.* New York: VCH, **1986**
- Nakanishi, K.; Berova, N.; Woody, R. W. (ed.) *Circular dichroism: Principles and applications.* New York: VCH, **1994**
- Richardson, F. S. *Theory of optical activity in the ligand-field transitions of chiral transition metal complexes. Chem. Rev.* **1979**, *79*, 17–36

General *LD* references

- Michl, J.; Thulstrup, E. W. *Spectroscopy with polarized light.* New York: VCH, **1986**
- Nordén, B. *Applications of linear dichroism spectroscopy. Appl. Spectrosc. Rev.* **1978**, *14*, 157–248
- Nordén, B.; Elvingson, C.; Jonsson, M.; Åkerman, B. *Microscopic behaviour of DNA during electrophoresis: electrophoretic orientation. Q . Rev. Biophys.* **1991**, *24*, 103–164
- Nordén, B.; Kubista, M.; Kuruscev, T. *Linear dichroism spectroscopy of nucleic acids. Q. Rev. Biophysics* **1992**, *25*, 51–170
- Schellman, J.; Jensen, H. P. *Optical spectroscopy of oriented molecules. Chem. Rev.* **1987**, *87*, 1359–1399
- Rodger, A. *Linear dichroism. Meth. Enzymol.* **1993**, *226*, 232–258

(1) Allen, R. E. (ed.) *The Concise Oxford Dictionary*; Oxford: Clarendon Press, **1990**
(2) Atkins, P. W. *Molecular Quantum Mechanics*; Oxford: Oxford University Press, **1983**
(3) Atkins, P. W. *Physical Chemistry*; 4th ed. Oxford: Oxford University Press, **1991**
(4) Hollas, J. M. *Modern Spectroscopy*; 2nd ed. Chichester: John Wiley and Sons, **1992**
(5) Davidsson, A.; Nordén, B. *Chemica Scripta* **1976**, *9*, 49; Nordén, B.; Seth, S. *Appl. Spectr.* **1985**, *39*, 647
(6) Nordén, B.; Kubista, M.; Kuruscev, T. *Q. Rev. Biophys.* **1992**, *25*, 51
(7) Rodger, A. *Meth. in Enzymol.* **1993**, *226*, 232
(8) Matsuoka, Y.; Nordén, B. *J. Phys. Chem.* **1982**, *86*, 1376; Nordén, B.; Seth, S. *Biopolymers* **1979**, *18*, 2323
(9) Abdourakhmanov, I. A.; Ganago, A. O.; Erokhin, Y. E.; Solov'ev, A. A.; Chugunov, V. A. *Biochim. Biophys. Acta* **1979**, *546*, 183
(10) Åkerman, B.; Johnsson, M.; Nordén, B. *Chem. Commu.* **1985**, 422
(11) Nordén, B.; Elvingson, C.; Jonsson, M.; Åkerman, B. *Quart. Rev. Biophys.* **1991**, *24*, 103
(12) Clark, L. B. *J. Phys. Chem.* **1990**, *94*, 2873
(13) Nordén, B.; Lindblom, G.; Jonás, I. *J. Phys. Chem.* **1977**, *81*, 2086

2 Circular dichroism of biomolecules

2.1 Introduction

It is probably true that *CD* is now more widely used to study biological systems than non-biological systems. This is despite the fact that the molecules concerned are usually very large and therefore detailed structural analysis of their *CD* data is not yet possible. In this chapter we shall focus on the *CD* of proteins, nucleic acids, and nucleic acid–ligand systems. What is written here is by no means the last word on these subjects as the increasing interest in structural details of biological molecules means that all techniques that can be used to study these systems are the focus of active research.

The *CD* of biological macromolecules is most commonly used in one of two ways:

- to probe changes in the conformation of the macromolecule itself, and
- to probe its interaction with small molecules, especially achiral ones whose induced *CD* is due solely to their interaction with the macromolecule.

Most of the data analysis that is performed for such systems is qualitative or empirical in nature. In this chapter we shall build on the general foundations laid in Chapter 1 to find out how to use *CD* to study biological systems.

There are many advantages of *CD* as a probe of biological molecules.

- The experiments are simple and quick to perform.
- *CD* is uniquely sensitive to the *asymmetry* of a system, and as the increasing emphasis on developing enantiomerically pure drugs shows, asymmetry can be a key feature of the interaction of a drug with its receptor.
- *CD* experiments deal with solution phase. This is important for biological systems because the crystallization process can change the structure of a molecule.[1,2]
- *CD* experiments require significantly lower concentrations than are appropriate for nuclear magnetic resonance (NMR). This means *CD* not only requires less sample but it can be used for systems where the concentrations required for NMR may change the system entirely.
- The intrinsic timescale of a *CD* measurement (10^{-15} s for the absorption of a UV photon) is defined by the energy of the photons used; it is thus much shorter than that for NMR which uses radio waves (10^{-9} s). This difference may be of relevance for macromolecule–

ligand interaction studies if the ligand has a particularly short residence time in a given orientation.

- A complete *CD* spectrum may be collected in a few minutes and single wavelength measurements require only a few milliseconds. It is generally the sample mixing process that defines the timescale of kinetic *CD* data, so *CD* has been widely used to probe protein folding using stopped-flow techniques.

- NMR and crystallography both give detailed structural information only for small proteins (molecular weight < 20,000) and nucleic acids (~24 bases) whereas *CD* can be used for any size of molecule.

Despite these advantages, *CD* can be considered under-utilized as a structural probe of biomolecules. The reasons for this probably include unfamiliarity with practical aspects of the technique as well as lack of basic information required for quantitative interpretation of experimental data. The difficulty of interpreting the spectra means that much of the structural information that the spectra contain is ignored.

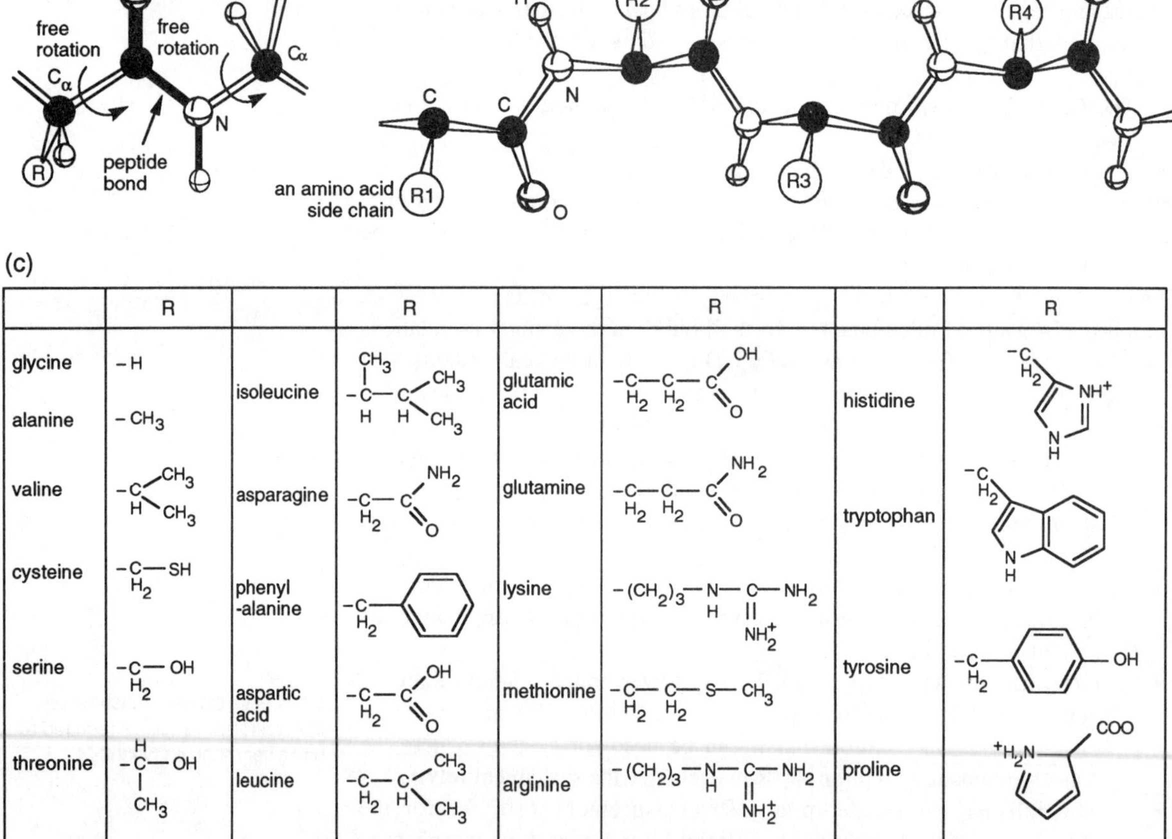

Fig. 2.1 (a) The peptide bond, (b) the primary structure of a protein, and (c) the commonly occurring L-amino acids.

2.2 *CD* of polypeptides and proteins

Protein geometry

Proteins are linear polymers of well-defined sequences of amino acids. Each amino acid is usually one of the twenty commonly occurring L-amino acids illustrated in Fig. 2.1; they are joined *via* what is known as the peptide bond which is formed between the acid group of one amino acid and the amine group of another by the elimination of a water molecule.

The protein primary structure (what is bonded to what) is illustrated in Fig. 2.1. One end of the protein molecule is thus the amino end (by convention this is the beginning) and the other end is the carboxylate end; these are usually referred to as the N terminal and C terminal respectively. The molecular identity of the protein is specified by listing its unique sequence of amino acids.

The overall shape of a protein molecule is crucial for its biological activity. In nature, proteins appear to adopt unique folded conformations that optimize this activity. Thus, any techniques that can provide information about the structure are potentially valuable to scientists endeavouring to understand how proteins function. Proteins form regular secondary structural units because the peptide unit O=C–N– is planar and rigid yet it has a large degree of rotational freedom about its links to the rest of the protein chain (Fig. 2.1a). This both constrains the possible relative orientations of neighbouring residues and allows a variety of possible intramolecular hydrogen bonding arrangements between the C–O of one peptide unit and the N–H of another unit.

Naturally occurring amino acids in most biological systems all have the same handedness [except for glycine which has R=H (Fig. 2.1) and is achiral]. As a result of the chirality of the amino residues, the secondary structure features thus formed are also chiral. The common structural motifs found in nearly all proteins are the α-helix (which for all L-amino acids is right-handed) and both parallel and anti-parallel β-sheets (Fig. 2.2). The turns between anti-parallel β-sheets are known as β-turns.

CD contains information about the asymmetric features of the backbone as well as about the orientations of the side chains. The challenge is to extract that information. Protein *CD* spectra are routinely empirically analysed by expressing them as a combination of standard spectra corresponding to a limited number of well-defined backbone geometries. The right-handed α-helix results when the nth peptide unit forms hydrogen bonds between its C–O and the N–H of the $(n + 4)$th peptide and between its N–H and the $(n - 4)$th C–O; there is a 1.5 Å translation and 100° turn between two consecutive peptide units, giving 3.6 amino acid residues per turn. An alternative efficient formation of hydrogen bonds occurs between a sheet of parallel or antiparallel runs of amino acids; these are known as a β-sheets (Fig. 2.2). Typically the strands of an anti-parallel β-sheet are linked by β-turns where the nth peptide unit forms hydrogen bonds with the $(n + 3)$rd peptide unit. If a β-sheet extends over more than two strands, then the relative arrangements of the strands in space must be considered.

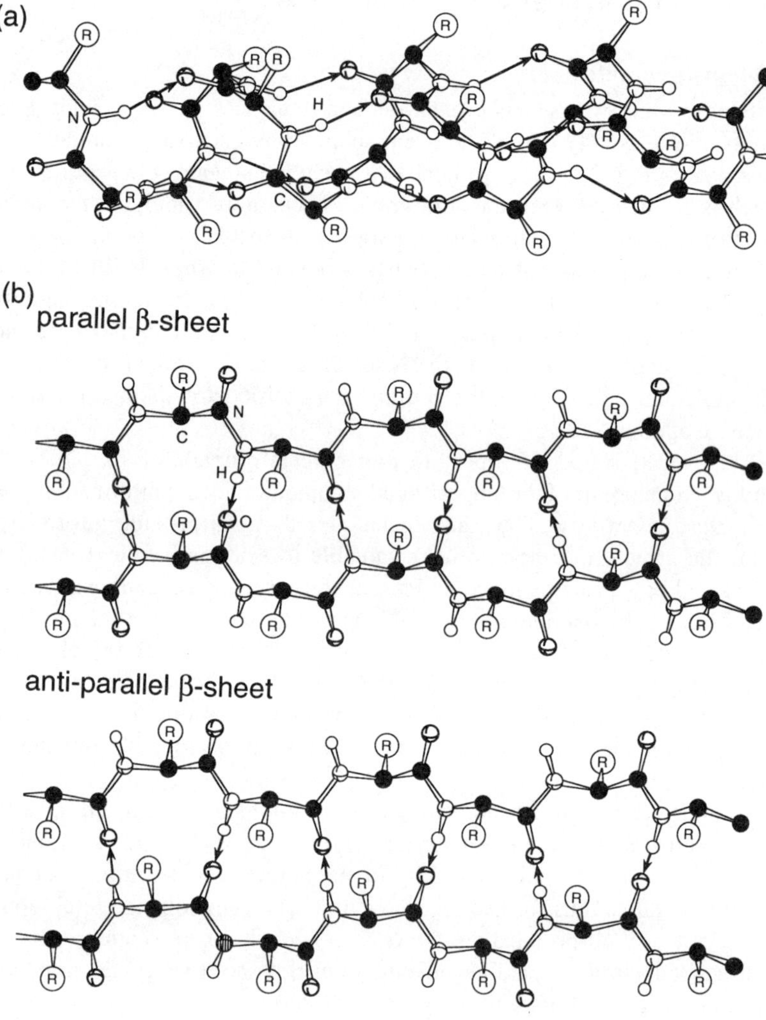

Fig. 2.2 (a) α-helix, and (b) β-sheet protein secondary structures. Arrows indicate H-bonds.

Protein UV spectroscopy

UV spectra of proteins are usually divided into the 'near' and 'far' UV regions. The near UV in this context means 250–300 nm and is also described as the aromatic region, though transitions of disulfide bonds (cystines) also contribute to the total absorption intensity in this region. The far UV (< 250 nm) is dominated by transitions of the peptide backbone of the protein, but transitions from some side chains also contribute in this region.

Some proteins, especially metallo-proteins, have so-called prosthetic or extrinsic groups with additional chromophores that can be used in analysis; the transitions often occur in the visible region and the groups can be analysed in the same way as the molecules that are discussed in Chapter 5.

The aromatic side chains, phenylalanine, tyrosine, and tryptophan all have transitions in the near UV region. The indole of tryptophan has two or more

It should be noted that if a protein has an extrinsic chromophore then that chromophore's spectroscopy may also interfere with the peptide region of the spectrum.

transitions in the 240–290 nm region with total maximum extinction coefficient ε_{max}(279 nm) ~5000 mol^{-1} dm^3 cm^{-1}; tyrosine has one transition with ε_{max}(274 nm) ~1400 mol^{-1} dm^3 cm^{-1}; phenylalanine also has one transition with ε_{max}(258 nm) ~190 mol^{-1} dm^3 cm^{-1}; and a cystine disulfide bond absorbs from 250 to 270 nm with ε_{max} ~300 mol^{-1} dm^3 cm^{-1}. Although tryptophans have by far the most intense transitions in this region, many proteins have few tryptophans compared with the other aromatic groups, so the region is not necessarily dominated by tryptophan transitions.

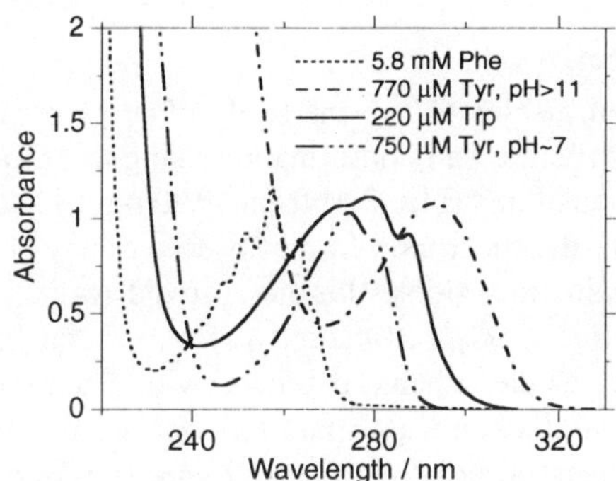

Fig. 2.3 Aromatic absorption spectrum of tryptophan, tyrosine, and phenylalanine.

The peptide chromophore (Fig. 2.4) which gives rise to the transitions observed in the far UV region (180–240 nm) has non-bonding electrons on the oxygen and also on the nitrogen atoms, π-electrons which are delocalized to some extent over the carbon, oxygen, and nitrogen atoms, and sigma bonding electrons.

The lowest energy transition of the peptide chromophore is an $n \rightarrow \pi^*$ transition analogous to that in ketones (Chapter 6), and the next transition is $\pi \rightarrow \pi^*$. As in the carbonyl case, the $n \rightarrow \pi^*$ transition is of low intensity (ε ~100 mol^{-1} dm^3 cm^{-1}), though it is not as low as for a simple ketone; it occurs at about 210–230 nm (depending mainly upon the extent of hydrogen bonding of the oxygen lone pairs) and its electric character is polarized more or less along the carbonyl bond. The $\pi \rightarrow \pi^*$ transition (ε ~7000 mol^{-1} dm^3 cm^{-1}) is dominated by the carbonyl π-bond and is also affected by the involvement of the nitrogen in the π orbitals; its electric dipole transition moment is polarized somewhere near the line between oxygen and nitrogen and it is centred at 190 nm. These transitions are schematically illustrated in Fig. 2.4.

As noted above, a number of amino acid side chains also have transitions in the peptide region.[3] Although these transitions are often stronger than the peptide $\pi \rightarrow \pi^*$ transitions, since the peptide chromophores are in excess the side chain transitions are generally nearly impossible to detect. However, the presence of these side chain transitions can be sufficient to confuse attempts to empirically determine the percentage of a given structural unit from *CD*. This is particularly true for proteins with a low α-helical content.

The chromophores with far UV transitions include the aromatic side chains, disulfide cystines, arginine, asparagine, aspartic acid, glutamine, glutamic acid, and histidine.

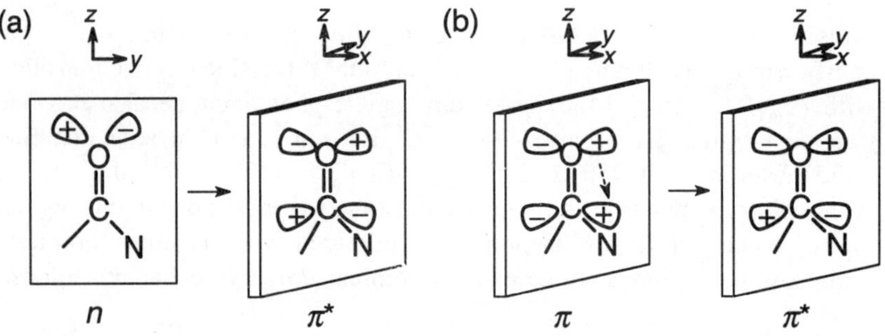

Fig. 2.4 Schematic illustrations of (a) $n \rightarrow \pi^*$ and (b) $\pi \rightarrow \pi^*$ transitions in peptides.

Protein *CD* spectra

At present the main use of *CD* in the study of proteins is as an empirical gauge of protein structure and conformation using the *CD* induced into the backbone amide transitions (Fig. 2.4) from ~190 nm to 240 nm (the far UV or peptide region of the spectrum). In the absence of any contributions to the *CD* from side chain transitions, this has proved to be a very successful approach. Distinctive *CD* spectra (Fig. 2.5) have been described for pure conformations such as the α-helix, β-sheets (with different ones sometimes being given for parallel and anti-parallel sheets), β-turns, and also the 'random' coil. At least in principle, the *CD* spectrum of a native protein is then the sum of the appropriate percentages of each component spectrum. As discussed below, there are a variety of methods for working backwards to determine the percentage α-helical *etc.* content of a given protein.

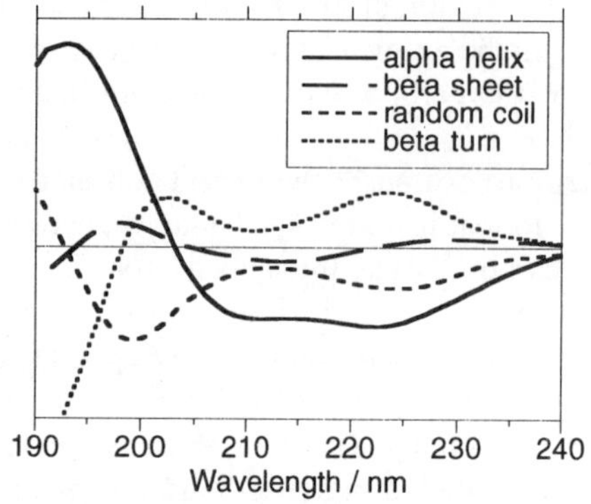

The fact that the random coil has a well-identified spectral form also requires it to have a well-defined structure, *i.e.* it is not random. In fact, many other structural units may also be identified, depending upon the detail with which crystal structures are examined.

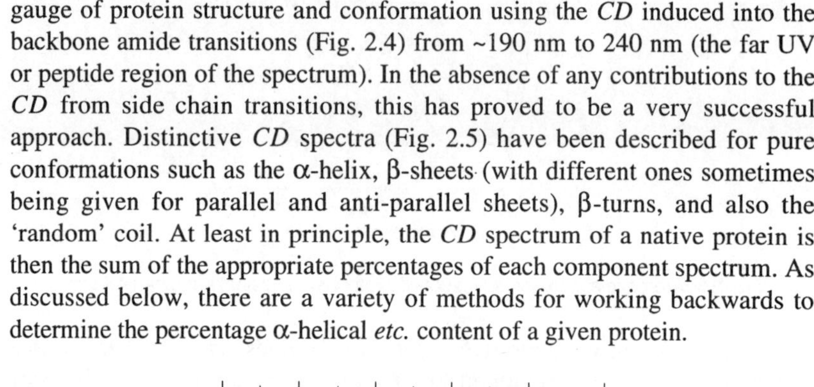

Fig. 2.5 Examples of standard spectra used in commercial *CD* protein structure fitting programs.

α-helix CD

The α-helix is the dominant secondary structure in many proteins and on average accounts for about one-third of the residues in globular proteins. It is a well-defined motif and is also adopted by many homopolypeptides under appropriate conditions (such as in trifluoroethanol). It has therefore been extensively studied. The *CD* spectra of α-helices are characterized by a negative peak with separate maxima of similar magnitude at 222 nm (the $n \rightarrow \pi^*$ transition) and 208 nm (part of the $\pi \rightarrow \pi^*$ transition).

The α-helix *CD* is larger in magnitude than that due to other motifs and has a distinctive spectral form at these wavelengths; it is apparent upon the most casual inspection of a spectrum. It is no surprise therefore that the various empirical *CD* fitting programs that are available are fairly successful in determining the percentage of α-helical content of a protein. It should, however, be noted that the magnitude of the *CD* signal does vary with variations in the helix and has been concluded to depend on helix length.

The length dependence of the α-helix *CD* is perhaps unsurprising: the signal behaves as if it were for a helix about four residues shorter than it in fact is; this corresponds to the number of unanchored hydrogen-bonding groups. In this context it should be noted that the description of a structural motif as α-helical when it is only about four amino acids in length is unhelpful. In such a case, the 'α-helix' spectrum will be dominated by 'end-effects'. It is largely for this reason that structural analysis of oligopeptides using *CD* has had problems.

The $\pi \rightarrow \pi^*$ transitions of the peptides in a polypeptide chain all couple together. For the right-handed α–helix of poly-L-lysine in aqueous solution the result is a negative signal at 205 nm for a transition whose polarization is parallel to the helix axis and a positive signal at 190 nm for a transition whose polarization is perpendicular to the helix axis. Other structural motifs show no such splitting of the $\pi \rightarrow \pi^*$ transition.[6]

β-sheet CD

The spectroscopic characterization of β-sheets has proved more difficult than that of α-helices due to the practical reason that they are less soluble in solvents with a good UV transmission and due to the intrinsic reason that they are generally structurally less well defined: they may be parallel or antiparallel and of varying lengths and widths. Furthermore, an extended β-sheet is usually found to show a marked twist, rather than to be planar. Such tertiary structure also influences the overall *CD* spectrum.

The general characteristics of β-sheet *CD* may be taken to be a negative band at about 216 nm and a positive band of comparable magnitude near 195 nm.[4] This level of characterization of the spectrum might be deemed to be not much worse than that of α-helices were it not for the fact that the 'random coil' spectral features have their maxima at similar wavelengths and are of opposite sign from those of the β-sheet. This means that an empirical fitting program may incorrectly weight these spectra (and the β-turn components) in an attempt to better account for the wavelength and magnitude variations that occur.

β-turn CD

The label β-turn is usually used to include all possible turns that occur, not simply the ones that enable a single strand to become an antiparallel β-sheet. About one-quarter of the residues in globular proteins then fall into this structural group. Despite this range of structures, a 'typical' β-turn *CD* spectrum has been identified which has a weak red-shifted negative $n \rightarrow \pi^*$ band near 225 nm, a strong positive $\pi \rightarrow \pi^*$ transition between 200 nm and 205 nm, and a strong negative band between 180 nm and 190 nm. However, the spectrum of a β-turn is not really as well defined as this characterization might imply.

Random coil CD

In an attempt to avoid the somewhat misleading implications of the label 'random coil', Woody has coined the phrase 'unordered conformation'.[5]

However, when we refer to random coils, we are generally grouping the parts of the folded protein that do not fit into one of the previously discussed categories which means that there may be ordered structures included in this 'residual'. The net *CD* of these parts of the protein has a strong negative *CD* signal just below 200 nm, a positive band at about 218 nm in many systems, and perhaps a very weak negative band at 235 nm.

Determining the percentage of different structural units in a protein

There are a range of different computer programs available for determining the percentage of different structural motifs just from the *CD* spectrum. The cost of these programs does not necessarily correlate with their reliability. All use 'standard' reference spectra such as those illustrated in Fig. 2.5 with respect to which a measured spectrum is decomposed so that an appropriately weighted sum of the reference spectra equals the measured spectrum. As the above description of the spectral features of different motifs indicates, data down to at least 190 nm should be used. However, there is the usual 'but'. The contributions to the observed *CD* from side chains can become significant below 200 nm, so limiting the fitting process to data above 200 nm wavelength may, in certain cases, be less misleading. It is currently accepted that α-helix estimates from most fitting programs will be reliable. Other estimates may need to be confirmed from independent sources.

Aromatic region

The intensity of the *CD* induced into the achiral aromatic side chains is very dependent on their environment so, in contrast to fluorescence, it is not possible with *CD* to limit consideration specifically to the tryptophans. Groups which can be essentially ignored in the absorption spectrum may become significant in the *CD*. In particular, disulfide groups (covalent bonds formed between cysteine residues in different parts of a protein) have transitions in the aromatic region of the spectrum. Although these transitions are weak (ε_{max} ~300 at 250 and 270 nm), they are magnetic dipole allowed and so contribute significantly to the protein *CD* in this region, often with tails stretching beyond 300 nm (which aids their identification).

As the side chains are usually isolated from one another analysing their *CD* usually becomes equivalent to the situation treated when we look at the *CD* induced into a ligand bound to the macromolecule. Their *CD* has proved to be particularly useful for probing conformational changes in a part of the protein containing one of these residues since the number of such chromophores is usually limited. The development of site-directed mutagenesis (whereby selected chromophores can be removed from the protein being studied) has greatly facilitated this approach.

Non-empirical analysis of protein CD

The important feature of the peptide $n \rightarrow \pi^*$ and $\pi \rightarrow \pi^*$ transitions compared with those of a simple carbonyl is that they occur very close in energy; the mixing of the transitions is therefore significantly enhanced compared with the situation in a carbonyl. All attempts to theoretically calculate peptide *CD* to date have foundered on this coupling.

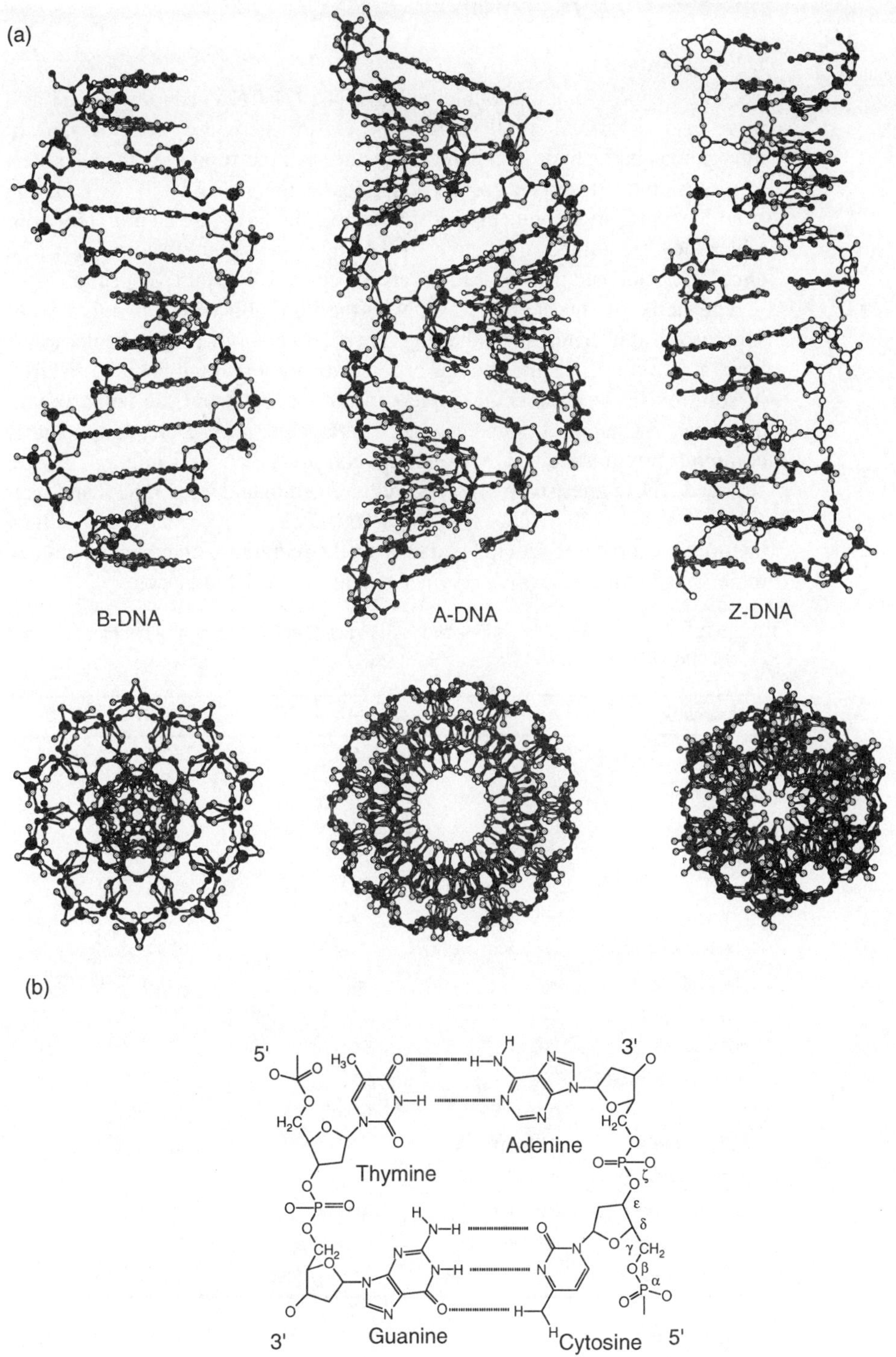

Fig. 2.6 (a) B-, A-, and Z-DNA viewed perpendicular to the helix axis and down the helix axis. (b) Structural formula of DNA illustrating the definitions of torsion angles.

2.3 *CD* of polynucleotides: DNA and RNA

DNA geometry

To a first approximation double helical DNA (or RNA) can be viewed as a more-or-less vertical spiral staircase where the steps are made of pairs of nitrogenous bases hydrogen bonded together and the support is the backbone of alternating phosphate groups and ribose sugars (Fig. 2.6). In isolation from the chiral sugar units the phosphates and the bases are achiral, however, when they are joined together with the sugar units they become part of a chiral molecule and their transitions are expected to exhibit *CD* spectra.

The helix of the standard B-DNA (the common solution polymorph) staircase is right-handed and the steps are *approximately* perpendicular to the helix axis. From *LD* studies it has become apparent that the bases of B-DNA in solution lie at an average angle of ~80° or less from the helix axis as discussed in Chapter 3. Some geometry parameters for B-DNA and the other common polymorphs A-DNA and Z-DNA are given in Table 2.1. Single stranded DNAs are structurally less well defined than duplex DNAs and their *CD* signal is smaller. *CD* spectra of RNAs (*e.g.* Fig. 5.2) clearly show naturally occurring RNAs to be mainly well structured with extensive duplex regions, rather than the often assumed single stranded geometry.

Table 2.1 Typical geometric parameters (see Figs 2.6 and 3.1 for definitions) for standard A-, B-, and Z-forms of DNA.[7–11]

Parameter	A-DNA	B-DNA	Z-DNA
α	−85°	−47°	60°/160°
β	−152°	−146°	−175°/−135°
γ	46°	36°	178°/57°
δ	83°	156°	140°/95°
ε	178°	155°	−95°/−110°
ζ	−46°	−95°	−35°–85°
sugar conformation	$C_{3'}$-endo	$C_{2'}$-endo	$C_{3'}$-endo/$C_{2'}$-endo
glycosidic bond	*anti*	*anti*	*anti* (C), *syn* (G)
base roll (twist)	12°	0°	1°
base tilt	20°	5°	9°
base helical twist	32°	36°	11°/50°
base slide	0.15 nm	0 nm	0.2 nm
helix diameter	2.55 nm	2.37 nm	1.84 nm
bases / turn of helix	11	10	12
base rise / base pair	0.23 nm	0.33 nm	0.38 nm
major groove	narrow, deep	wide, deep	flat
minor groove	broad, shallow	narrow, deep	narrow, deep

RNA differs from DNA in that every T is replaced by U, which has one less methyl group, and the backbone sugar is ribose instead of deoxyribose. Double helical RNA adopts the A-form geometry. The repeat unit of Z-DNA is a dinucleotide, necessitating two values of each angular parameter. ζ adopts a range of values in Z-DNA.

UV spectroscopy of the DNA bases

The UV absorbance of nucleic acids from 200 to 300 nm is due exclusively to transitions of the planar purine and pyrimidine bases. The backbone begins to contribute at about 190 nm. The accessible region of the spectrum is therefore dominated by $\pi \to \pi^*$ transitions whose exact assignment is still a matter of debate. All of them are polarized in the plane of the bases. The proposed assignments for guanine and cytosine are illustrated in Fig. 2.7a. The UV spectra of the bases (Fig. 2.7b) look as if there are two simple bands, however, each 'simple' band observed is generally a composite of more than one transition.

DNA *CD*

When we measure a *CD* spectrum we are probing the asymmetry of the system. If we consider an isolated nucleotide, the chirality originates from the ribose sugar. In the spectral region to which we have relatively easy access (down to 180 nm), the chiral ribose-phosphate backbone has no important transitions. Therefore, when we measure the *CD* spectrum, we are detecting the *CD* induced into the transitions of the base as a result of their coupling with the backbone transitions. The magnitude of $\Delta\varepsilon_{max}$ is of the order of 2 mol^{-1} dm^3 cm^{-1} at 270 nm and the purine bases have a negative signal whereas the pyrimidine ones have a positive signal (Fig. 2.8).

There may be some $n \to \pi^*$ transitions present whose intensity (even in *CD*) is so small that they have not been definitively identified.

A nucleotide is a base plus sugar plus phosphate; a nucleoside is a base plus sugar.

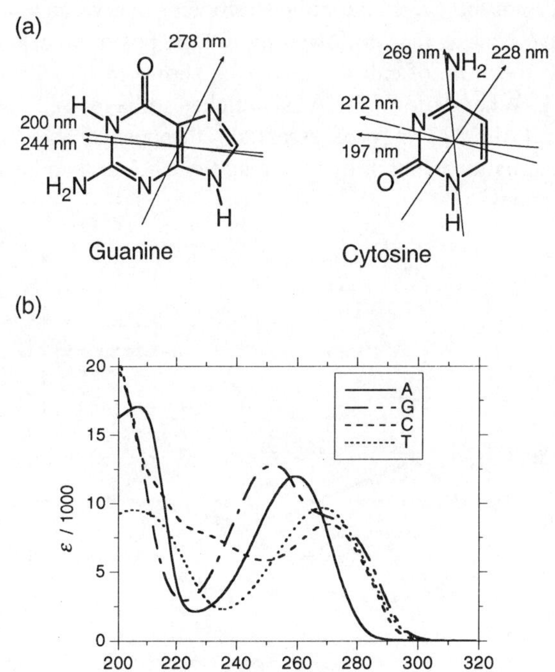

Fig. 2.7 (a) Probable transition polarizations for UV transitions of guanine and cytosine.[12] (b) UV spectra of the DNA nucelotides (A) deoxyadenosine 5'-monophosphate, (G) deoxyguanosine 5'-monophosphate,(C) deoxycytidine 5'-monophosphate, and (T) thymidine 5'-monophosphate (that of uracil is very similar).

In the case of DNA and RNA polymers it is still the sugar units of the backbone that provide the orginal chirality. However, when we measure a *CD* spectrum of a polynucleotide with stacked bases, the magnitude is larger at 270 nm and significantly larger at 200 nm than that of the individual bases. What we are measuring is dominated by the *CD* induced into transitions of the bases from their coupling with each other since the bases stack on top of each other in a chiral (helical) fashion.

Empirical structural analysis of DNA CD

The simplest application of *CD* to DNA structure determination is for identification of which polymorph we have in our sample. Since the DNA *CD* from 200 to 300 nm is due to the skewed orientation of the bases, if the DNA is untwisted or the bases are tilted a change from what is observed for B-DNA would be expected in the *CD* spectrum. Somewhat surprisingly, observed *CD* spectra often vary more with changes in base orientation (DNA polymorph) than as a function of the base composition of the DNA, though they are of course a sensitive function of DNA sequence as well.

The *CD* signature of B-form DNA (Figs 2.9 and 2.10) as read from longer to shorter wavelength is a positive band centred at 275 nm, a negative band at 240 nm, with the zero being around 258 nm. These two bands are not from a simple degenerate exciton coupling (see Chapter 5), but rather the superimposed result of all the couplings of the transitions present in all the bases. At 220 nm the *CD* signal either becomes positive or less negative; a small negative peak is then followed by a large positive peak from 180 to 190 nm. The spectrum of calf thymus DNA shown in Fig. 2.9 is typical of B-form DNA. When B-form DNA is stretched to have 10.2 bases per turn instead of 10.4, the 275 nm peak essentially disappears. This form of DNA can be induced by methanol or by wrapping the DNA around histone cores to form nucleosomes.

The couplings are short range, so only neighbouring bases have a large effect on each other.

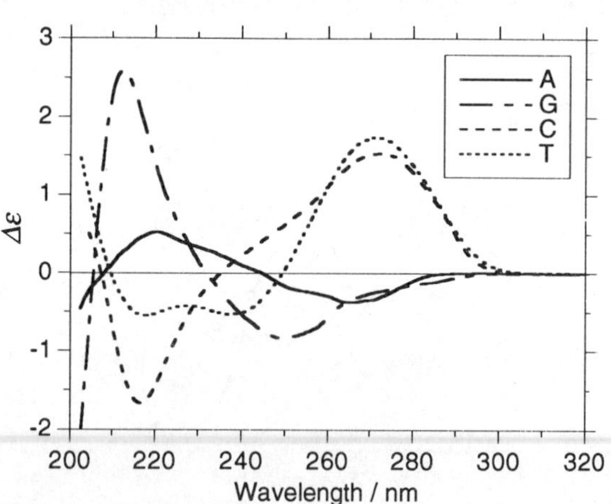

Fig. 2.8 *CD* spectra of isolated DNA nucleotides: (A) deoxyadenosine 5'-monophosphate, (G) guanosine 5'-monophosphate, (C) deoxycytidine 5'-monophosphate, and (T) thymidine 5'-monophosphate (that of uracil is very similar).

If B-DNA is compacted and the bases tilted and radially displaced from the centre of the helix (thus creating a hole when one looks down the helix axis) (Fig. 2.6 and Table 2.1) then A-form DNA results. A-DNA is characterized by a positive *CD* band centred at 260 nm that is larger than the corresponding B-DNA band, a fairly intense negative band at 210 nm and a very intense positive band at 190 nm. The 250–230 nm region is also usually fairly flat though not necessarily zero. Naturally occurring RNAs adopt the A-form if they are duplex. A typical natural RNA spectrum is shown in Fig. 5.2.

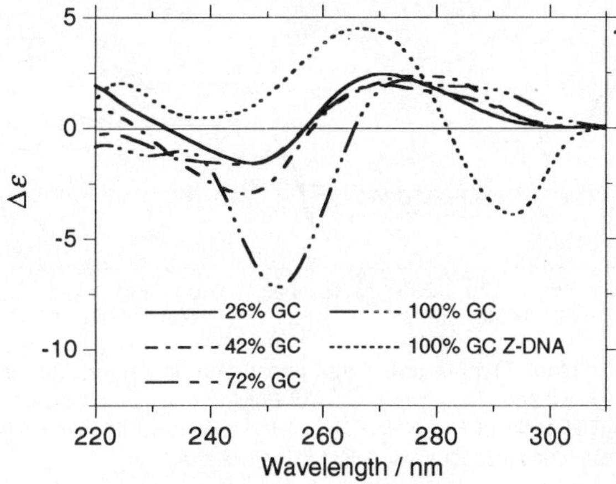

Fig. 2.9 *CD* spectra of duplex DNAs (pH = 6.8, 5 mM NaCl, 1 mM cacodylate) of varying G-C base pair content: *Clostridium perfringens* (27% GC), calf thymus (42% GC), *Micrococcus lysodeikticus* (72% GC), poly[d(G-C)]$_2$ (100% GC). Also shown is Z-form poly[d(G-C)]$_2$ (100% GC, pH = 6.8, 5 mM NaCl, 1 mM cacodylate, 50 µM [Co(NH$_3$)$_6$]$^{3+}$).

Z-form DNA (Fig. 2.6 and Table 2.1) does not readily form for all sequences. However, it is easily formed for poly[d(G-C)]$_2$ in the presence of highly charged ions. It is characterized by a negative *CD* band at 290 nm and a positive band at 260 nm. Care must be taken in using these signatures to identify Z-form DNA, since the same DNA in the A-form has a negative band at 295 nm and a positive band at 270 nm.[13] A more definitive signal is the large negative *CD* signal in the 195–200 nm region for Z-DNA, whereas B-form DNA *CD* is near zero or positive in this region. For Z-DNA the *CD* passes through zero between 180 and 185 nm.

A *CD* spectrometer is also an ideal tool for probing condensed DNA structures, although what is measured then is not strictly *CD* but rather a form of light scattering. The results of intramolecular condensation are particles of size comparable with the wavelength of the light. The particles interact differently with left and right circularly polarized light and, at the wavelengths where the chromophores absorb light, give a large '*CD*' signal, the sign of which tells the helical handedness of the condensed DNA particle, with a negative signal above 250 nm (Fig. 2.10a) corresponding to a left-handed helix.[14]

CD spectra are so easy to measure that it is often the simplest technique to use to probe DNA conformational changes as a function of ionic strength,

The 195 nm negative signal of Z-DNA is obscured by the *CD* induced into the transitions of [Co(NH$_3$)]$^{3+}$ when this molecule is used to induce Z-form DNA.

solvent, ligand concentration, *etc.* A titration showing the B to Z transition of poly[d(G-C)]$_2$ as spermine is added is shown in Fig. 2.10b. Although absorption is more commonly used to monitor the helix (double stranded DNA) to coil (single stranded DNA) transition of DNA, *CD* is actually more sensitive because of its sensitivity to small changes in base stacking. In this way, the conformation of DNA at different temperatures may be identified.

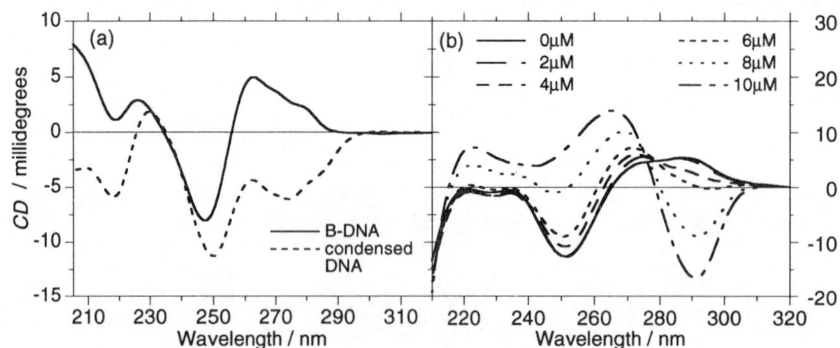

Fig. 2.10 (a) Poly[d(A-T)$_2$] (44 µM, 1 mM phosphate) in B-form (20 mM NaCl) and condensed (1 M NaCl and 45% v/v ethanol). (b) Poly[d(G-C)$_2$] as a function of increasing spermine, [NH$_3$(CH$_2$)$_3$NH$_2$(CH$_2$)$_4$NH$_2$(CH$_2$)$_3$NH$_3$]$^{4+}$, concentration showing the B → Z transition. Spermine concentrations are indicated in the figure.

Quantitative structural analysis of DNA CD

With DNA, as discussed above, it is usually possible to qualitatively identify the overall polymorph that is present, but a more quantitative *empirical* analysis has defied extensive efforts. Somewhat perversely, however, theoretical analysis of DNA *CD* performed by calculating the spectrum for a proposed geometry is possible, whereas for proteins this approach has been less successful.

The *CD* spectrum measured for a solution of DNA molecules is sensitively dependent upon both the sequence and the conformation of the DNA. It is therefore a most attractive proposition to be able to analyse the details of sequence-dependent DNA structure using *CD*. Being realistic, we should acknowledge that this will only be possible for comparatively short DNAs or long ones that are multiple repeats of a short sequence.

On the assumption that magnetic dipole allowed $n \rightarrow \pi^*$ transitions may be ignored, then we can conclude that the *CD* arises from coupled oscillator-type couplings of degenerate and near-degenerate oscillators in the bases as is discussed in Chapter 5. The simple case of a dinucleotide with one transition on each base is covered in §5.7.

The method used to perform *CD* calculations for DNA has been developed by Tinoco, Schellman and others.[13,15] The label 'matrix method' has been coined because the key computational step involves the diagonalization of an interaction matrix. Instead of using transition dipole vectors as input for the exciton calculations described in Chapter 5, in the matrix method the transitions are expressed as collections of transition monopoles—with the transition monopoles being the changes in charge occurring at various

When the twist direction is reversed from B-DNA to Z-DNA, an inversion of the *CD* spectrum might be expected. While this is indeed true to some extent, we must not assume that B-DNA (which has parallel bases and a right-handed twist) and Z-DNA (which also has parallel bases and a left-handed twist) have equal and opposite *CD*s. This would only be true if they were actual mirror images of one another. What has happened in going from B to Z is that the *average* skew angle has (more or less) reversed, but not the skew angle of any particular pair of coupling dipoles. The intrinsic chirality of the backbone ribose units is of course the reason for this lack of reflection symmetry.

points, usually the atoms, during the transition. The couplings between the transitions are determined from the monopole description of the transition, and the resulting *CD* then calculated. In practice, it may be convenient to also include a dipole description of the transition, but if the monopole description is accurate this can be determined from it.

As DNA bases are really too close together for the multipole expansion approach used in Chapter 5 to be entirely valid the alternative used in the matrix method involves distributed monopoles instead. In general, as input for the calculation of DNA *CD* we require nucleo-base transition moments, transition monopoles (which can be obtained from quantum mechanical calculations), transition energies, and the geometry of the DNA. If these are correct then a calculation should reproduce the experimental spectrum. If they are incorrect, then they cannot. To date, the experimental *CD* spectrum of poly[d(G-C)]$_2$ DNA has been reproduced reasonably well computationally. However, less success has been had with sequences containing adenines and thymines. Part of this problem has undoubtedly been the uncertainty in transition moment polarizations, however, at least as great a source of error may be that the standard coordinates for A-T rich sequences do not reflect the solution geometry very well. Fortunately, this is no longer the problem it was as molecular modelling techniques have been developed to the stage where fully solvated DNA simulations with reasonable potentials can be performed. It should be noted, however, that it is not sufficient to take a single DNA geometry from a molecular dynamics trajectory since the variation in geometry as a function of time is large. The way in which *CD* can potentially be used as a detailed probe of DNA structure is therefore illustrated in Fig. 2.11.

The transition monopoles should comply with the experimental transition moments.

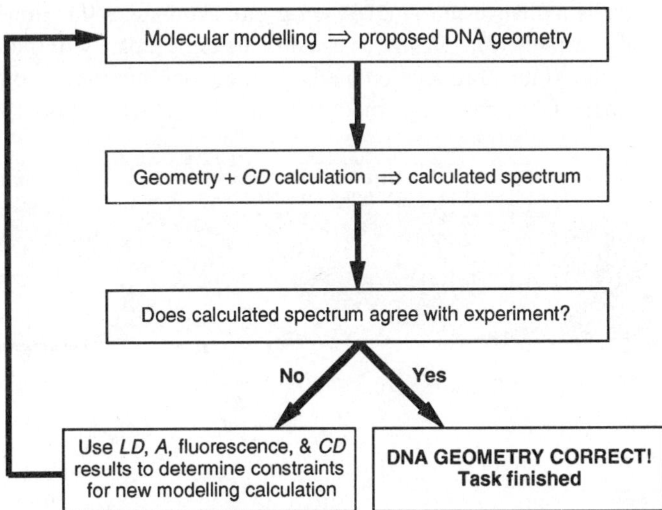

Fig. 2.11 Schematic illustration of a self-consistent method for determining DNA structure from *CD* data by coupling calculations with molecular modelling studies of DNA systems.

2.4 DNA–ligand interactions

If the ligand transitions of interest occur in the DNA region of the spectrum we take the *ICD* to be the total *CD* of the system minus that of the DNA at the same concentration. This means any DNA conformational changes that alter the DNA *CD* are included in the *ICD*.

Many DNA binding molecules (usually referred to as ligands or drugs even when they may have little or no therapeutic value) are themselves achiral. However, when they bind to DNA they acquire an induced *CD* (*ICD*) that is characteristic of their interaction. As with application of *CD* to proteins and DNA the ligand *ICD* can be used on a number of levels. The simplest use is to note that it exists and therefore conclude that the molecule *does* bind to DNA. However, more information can usually be extracted.

Empirical analyses of ligand ICD

It is easiest to deduce the *CD* changes per ligand if the ligand concentration is kept constant. Some ways of doing this were given in §1.3.

It is often very useful to measure a series of spectra where some variable such as the ionic strength, or mixing ratio (drug : DNA ratio), or temperature is changed. If the *CD* intensity changes but the shape of the spectrum remains the same during such an experiment, then it can be deduced that the ligand binding mode is unchanged though the amount of bound ligand may have changed (*e.g.* Fig. 2.12a).

If the ligand binds in a single binding mode, or in a number of sites whose relative proportions are independent of DNA:drug ratio, then we may write

$$ICD = \frac{L_b}{\alpha} \tag{2.1}$$

where L_b is the concentration of bound ligand, and α is a proportionality constant. For very high DNA concentrations the total ligand concentration will approximately equal the bound ligand concentration. Some of the simpler analysis methods that may be used to determine binding constants and site sizes are outlined in Appendix 2.

If, however, there is a change in the shape of the spectrum (Fig. 2.12b) as the mixing ratio or other experimental variable is changed, this certainly implies there is a change in the DNA–drug interaction as a function of the experimental variable. The change is usually due to occupancy of more than one binding site as the drug load on the DNA increases, but may also be due to changes in the DNA conformation or to ligand–ligand interactions.

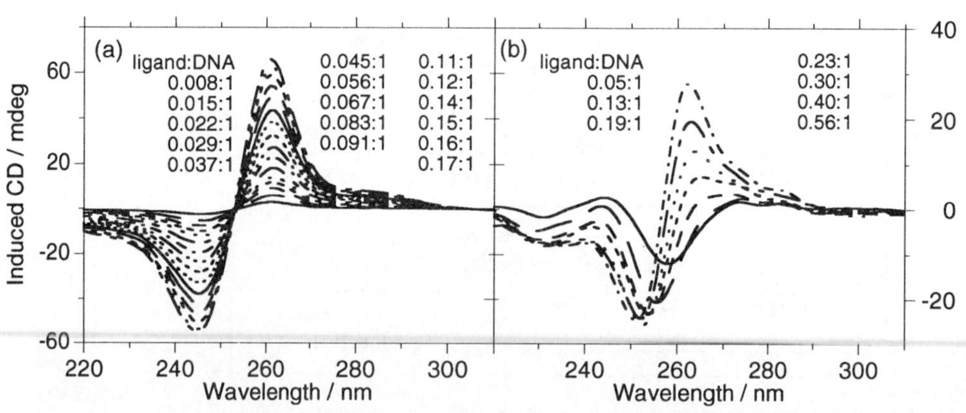

Fig. 2.12 *ICD* of varying concentrations of anthracene-9-carbonyl-N^1-spermine upon interaction with (a) 160 µM poly[d(G-C)$_2$], 5 mM NaCl, and low ligand : DNA phosphate mixing ratios, and (b) 130 µM poly[d(A–T)$_2$], 5 mM NaCl, and a wide range of ligand : DNA : phosphate mixing ratios. $\ell = 1$ cm.[16] The smallest magnitude *ICD* signal corresponds to the lowest ligand concentration.

If the experiment is conducted at constant ligand concentration and varying DNA concentration, then unless the ligand binding mode is changing there should be no change in the observed ligand *ICD*. A particularly dramatic change is observed, if as the DNA concentration is *decreased* (so one would expect fewer ligands to bind and so a decrease in *ICD*), the ligands begin to stack together and the *ICD* increases. Under these circumstances, we generally observe a large excitonic *CD* (Chapter 5). Thus spectra such as those of Fig. 2.13 immediately tell us that significant ligand–ligand interactions are occurring on the DNA.

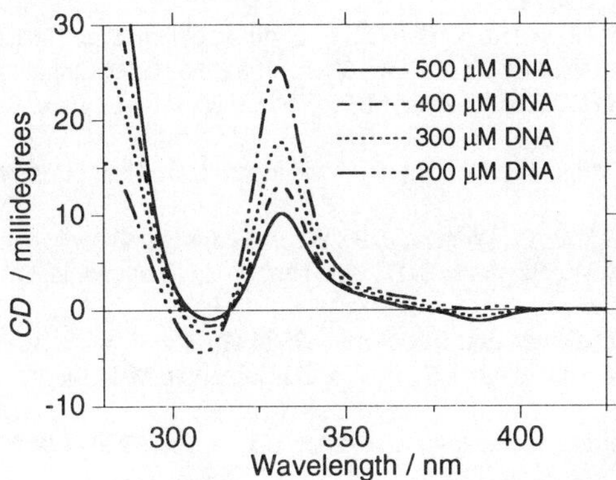

Fig. 2.13 *CD* of constant concentration of 9-hydroxyellipticene (50 µM) in the presence of calf thymus DNA (concentrations indicated in figure, 5 mM NaCl, 1 mM phosphate buffer).

Analytic use of ligand *CD*

The *CD* induced into ligand transitions upon binding to DNA almost always arises as a result of non-degenerate exciton coupling (see Chapter 5) between the electric dipole transition moment of the ligand and those of the DNA bases. The simple exciton approach of Chapter 5 can be used to derive an equation for the net effect of all the bases in a piece of DNA on a given ligand transition. This particular example is pursued in more detail in §5.7 for the case where the ligand is intercalated between two DNA base pairs.

For other types of binding, although we can write equations describing their coupled oscillator *CD*,[17] they have not been particularly easy to use. It is anticipated that more success will be had by following the programme outlined in Fig. 2.11. The exciton *CD* arising from the coupling of ligands stacked on DNA can be interpreted using the degenerate exciton *CD* of Chapter 5 to give the relative orientations of the ligands on the DNA.

References

• For general references, see Chapter 1.

(1) Benevides, J. M.; Wang, A. H.-J.; Marel, G. A.; van Boom, J. H.; Thomas, G. J. *Biochemistry* **1988**, *27*, 931

(2) Lipanov, A.; Kopka, M. L.; Kaczor-Grzeskowiak, M.; Quintana, J.; Dickerson, R. E. *Biochemistry* **1993**, *32*, 1373

(3)   Woody, R. W. *Biopolymers* **1978**, *17*, 1451

(4)   Greenfield, N.; Fasman, G. D. *Biochemistry* **1969**, *8*, 4108

(5)   Woody, R. W. *Circular dichroism of peptides and proteins* in *Circular dichroism principles and applications*, Nakanishi, K.; Berova, N.; Woody, R. W. (eds) New York: VCH, **1994**

(6)   Rosenbeck, K.; Doty, P. *Proc. Nat. Acad. Sci. USA* **1961**, *47*, 1775

(7)   Egli, M.; Williams, L. D.; Gao, Q.; Rich, A. *Biochemistry* **1991**, *30*, 11388

(8)   Stryer, L. *Biochemistry*; 3rd ed.; New York:W.H. Freeman and Company,**1988**

(9)   Calladine, C. R.; Drew, H. R. *Understanding DNA, The molecule and how it works*; Cambridge: Academic Press Ltd., **1992**

(10)   Beveridge, D. L.; Jørgensen, W. L. *Ann. NY Acad . Sci.***1986**, 482

(11)   Gessner, R. V.; Frederick, C. A.; Quigley, G. J.; Rich, A.; Wang, A. H.-J. *J. Biol. Chem.* **1989**, *264*, 7921

(12)   Zaloudek, F.; Novros, J. S.; Clark, L. B. *J. Amer. Chem. Soc.* **1985**, *107*, 7344; Matsuoka, Y.; Nordén, B. *J. Phys. Chem.* **1982**, *86*, 1378; Clark, L. B. *J. Am. Chem. Soc.* **1994**, *116*, 5265; Fülscher, M. P.; Roos, B. O. *J. Am. Chem. Soc.* **1995**, *117*, 2089

(13)   Refs. in Williams, A. L. Jr.; Cheong, C.; Tinoco, I. Jr.; Clark, L.B. *Nucl. Acids Res.* **1986**, *14*, 6649

(14)   Arscott, P. G.; Ma, C.; Wenner, J. R.; Bloomfield, V. A. *Biopolymers*, **1995**, *36*, 345; Patterson, C. W.; Singham, S. B.; Salzman, G. C.; Bustamente, C. *J. Chem. Phys.* **1986**, *84*, 1916

(15)   Rizzo, V.; Schellman, J.A. *Biopolymers* **1984**, *23*, 435

(16)   Rodger, A.; Blagborough, I. S.; Adlam, G.; Carpenter, M. L. *Biopolymers* **1994**, *34*, 1583

(17)   Hiort, C.; Nordén, B.; Rodger, A. *J. Amer. Chem. Soc.* **1990**, *112*, 1971

3 Linear dichroism of biomolecules

3.1 Introduction

In this chapter we shall see how LD can be used to study macromolecules of biological interest. In contrast to the previous chapter, we shall deal with proteins fairly quickly as LD has been less widely used for these molecules than for DNAs. Our main concern will be with DNA since LD, particularly flow and electric field LD, has been used to great effect in the study of the conformation of DNA and binding geometries of DNA–drug systems. Since LD is conceptually quite simple (much simpler than CD) the reader will be able to follow easily the deduction of structural information for biopolymers in the examples of this chapter. The formal mathematical framework for LD is presented in Chapter 4 for more general cases of LD spectroscopic applications.

Qualitative information about how the light-absorbing chromophores are oriented in space may be obtained simply from the sign of the LD, as was discussed in §1.2. For many situations the following formula (derived in Chapter 4) also allows a quantitative characterization of molecular orientation or structure to be performed:

$$LD^r = \frac{LD}{A} = \frac{A_{\parallel} - A_{\perp}}{A} = \frac{3}{2}S\left(3\cos^2\alpha - 1\right) \tag{3.1}$$

in which LD^r is the so-called reduced LD, S is a scaling factor (the orientation factor) defining the efficiency of macroscopic orientation, and α is an angle that specifes the orientation of the transition moment that is responsible for the absorption of light at the particular wavelength (if several transitions absorb at the same wavelength an average is obtained, see Chapter 4). In eqn (3.1) A denotes the absorption of the same or a corresponding sample under isotropic, *i.e.* unoriented, conditions. Note that A refèrs to the same concentration, C, and sample thickness, ℓ, as the LD. Thus, $A_{\parallel}$, $A_{\perp}$, and A all have $C\ell$ as a common factor, hence we do not need to know the sample concentration and pathlength.

Eqn (3.1) holds for any oriented system in which the transition moment is directed at an angle α relative to a local reference (or orientation) axis in such a way that all orientations on a cone around this reference axis are statistically equally probable. This is for example the case with helical structures in which chromophoric units are wound around a central helix axis when one passes from one unit to the other in the structure. The condition of uniform distribution around the reference axis may also be fulfilled for

$S = 1$ for perfect orientation and $S = 0$ for random (*i.e.* no) orientation.

Local reference axes include the normal to a membrane surface, the helix axis of a protein α-helix, or the helix axis of a DNA duplex.

dynamic reasons, such as by free rotation about the backbone of a linear polymer molecule.

In cases where the sample also fulfils the requirement of being macroscopically uniaxial, such as a polymer film drawn in one direction or polar molecules in an electric field, there is a simple relation beween $A_{\parallel}$, $A_{\perp}$, and A which makes it unnecessary to measure all three quantities:

$$A = \frac{1}{3}\left(A_{\parallel} + 2A_{\perp}\right) \tag{3.2}$$

3.2 Proteins

Attempts to use *LD* to study protein systems usually encounter one of two problems: either the system is difficult to orient (as is certainly the case for globular proteins); or, once oriented, the structure surrounding the oriented protein, such as a cell membrane or a filamentous aggregate, that is required to create orientation may give rise to artefacts caused by light scattering. These problems are, however, generally surmountable and *LD* has been used to good effect, particularly in the photosynthesis field, to determine orientations of chromophores, mobility of particular centres, *etc.* and for studying DNA–protein interactions.[1]

LD of polypeptides

The UV spectroscopy of amino acids and proteins was outlined in §2.2. Typical absorption spectra of peptides in the 180–240 nm region show features due to three electronic transitions: two stronger exciton (see Chapter 5) transition components of the $\pi \rightarrow \pi^*$ amide transitions of each residue centred around 200 nm and the weaker $n \rightarrow \pi^*$ transitions at about 220 nm. As discussed in §5.5, the two net $\pi \rightarrow \pi^*$ transitions of a peptide should have electric transition moments, $\boldsymbol{\mu}$, that are polarized at right angles to each other. For a polypeptide α-helix, one component (at 210 nm) is polarized along the helix axis and the other (at 190 nm) is polarized perpendicular to the helix axis. If the $n \rightarrow \pi^*$ transition belonged within a pure carbonyl chromophore (see Chapter 6), it would be electric dipole forbidden. For peptides, however, the lower symmetry of the amide chromophore enables some coupling with the π system and the transition acquires some electric character and an electric transition moment on each amide (Fig. 2.4).

Let us now consider the polarized spectra, $A_{\parallel}$ and $A_{\perp}$, that might be recorded using a thin humid film of poly-L-glutamic acid oriented by stroking on a silica plate. The parallel of $A_{\parallel}$ means that the light is polarized parallel to the orientation direction of the film, which should also be the direction of preferred orientation (helix axis) of the α-helical polypeptide filaments. Some typical data might be

Wavelength / nm	190	200	210	220	230	240
$A_{\parallel}$	0.45	0.55	0.56	0.38	0.22	0.18
$A_{\perp}$	0.63	0.55	0.47	0.38	0.27	0.18

If we then calculate the *LD* using eqn (3.1) and the isotropic absorbance using eqn (3.2), we may determine the concentration- and pathlength-independent reduced *LD*, LD^r, as follows.

Wavelength (nm)	190	200	210	220	230	240
LD	−0.18	0	0.09	0	−0.05	0
A	0.57	0.55	0.50	0.38	0.25	0.18
LD^r	−0.32	0	0.18	0	−0.20	0

The negative and positive *LD* signs at 190 and 210 nm are consistent with absorptions that are polarized, respectively, perpendicular and parallel to the fibre axis (*cf.* Fig. 5.11). The negative sign of the 230 nm band indicates that the $n \rightarrow \pi^*$ transition is also preferentially perpendicularly polarized. If we assume that the absorption at 190 nm is perpendicularly polarized with $\alpha = 90°$, we may conclude that the orientation factor *S* of the system is 0.21. Deviation from perfect perpendicular polarization ($\alpha < 90°$) for the $\pi \rightarrow \pi^*$ transition would result in our calculated value of *S* being too low.

Overlap of transitions of different polarizations also leads to calculated values of *S* being too low. For example, if the intensity at 210 nm is due to a parallel polarized transition with no overlap by transitions of other polarizations then $\alpha = 0°$ and substitution of the LD^r value of 0.18 into eqn (3.1) would give the value of *S* to be only 0.09. This indicates that there is a substantial overlap from the surrounding oppositely polarized transitions. The 210 nm transition is, in fact, only seen as a shoulder in the absorption spectrum and is thus strongly overlapped by the ones at higher and lower energies which are of different polarization. Transition overlap is less of a problem for the $n \rightarrow \pi^*$ and higher energy component of the $\pi \rightarrow \pi^*$ transitions, thus explaining the diverging results. The more general equations derived in Chapter 4 allow for determination of the orientation factor, polarization angles, and the extent of absorption overlap.

Membrane proteins

Membrane proteins are naturally oriented systems so *LD* is in principle an ideal technique to study them. An orientation technique currently used is squeezed gels (*cf.* §1.4). For example, Griebenow *et al.* have used this method to deduce the relative orientation of bacteriochlorophyll transition moments relative to the long axis of the chlorosomes.[2] In order to analyse experimental *LD* data quantitatively it is important to determine *S* [eqn (3.1)]. In the absence of a definitive value, *S* has often been assumed to be 1 in gel oriented protein studies, which leads to the final results being suspect [*e.g.* a low value of LD^r and ignoring *S* in eqn (3.1), results in a virtual value of α close to the magic angle 54.7°]. An attempt to determine *S* for a gel oriented DNA–protein complex is given in reference 3.

Single crystal orientation of photochemical reaction centres has also been used to determine orientations of chromophores[3] and Wan *et al.* have used time-dependent *LD* studies to probe the conformational motion in bacteriorhodopsin over the first 10 μs after excitation.[4] In principle membrane proteins could be studied directly in their anisotropic membrane, however, experimental difficulties arise because the optical axis of the sample (which is normally a thin layer through which the light beam is propagated) aligns with the propagation direction of the light. Nordén *et al.* have developed a method to study uniaxial samples where the optic axis is perpendicular to the plane of the sample by inclining the optical axis to the direction of the incident radiation.[5]

DNA–protein interactions

Flow *LD* is particularly suited to the study of non-specific DNA–protein interactions. *LD* titrations can be used to determine the amount of complex present since the protein changes the orientation properties of the DNA markedly. Further, changes in overall spectra and shape indicate formation of new complexes. Electric field orientation and gel techniques have also been used to study DNA–protein complexes, though measurements with the latter systems are restricted to wavelengths greater than 250 nm.[2]

Two DNA–protein systems that have been extensively studied are chromatin (in which DNA is both wrapped round the histone proteins to form beads and also links the beads together) and DNA complexes with the recombination enzyme *Rec*A. An interesting warning arises from some of the work on chromatin: inconsistent data from different orientation techniques may simply reflect different orientations, rather than any perturbation of the molecular structure by the applied fields. This appears to be the case for chromatin partly condensed by Mg^{2+} ions when oriented by flow or by an electric field. In the former case, the orientation is determined by the whole of the chromatin, whereas the electric field orientation is dominated by the more polarizable counter ion clouds of the linker regions of DNA.[6]

*Rec*A binds extremely cooperatively to DNA and thereby also strongly enhances the hydrodynamic orientation of the DNA providing information both about the binding geometry and about the binding rate and stoichiometry.[7] In another study, one of the two *Rec*A tryptophans was replaced by threonine which is transparent in the tryptophan region.[8] As the binding of the wild type and the mutant *Rec*A to DNA was the same, difference spectroscopy was used to enable the orientation of the tryptophan to be determined.

3.3 Nucleic acids

The geometry and UV spectroscopy of nucleic acids was discussed in §2.3. The absorption, and hence *LD*, spectra of nucleic acids in the easily accessible UV region of the spectrum (down to 180–190 nm) are dominated by the $\pi \rightarrow \pi^*$ transitions of the purine and pyrimidine bases, although there is some evidence for weak out-of-plane polarized $n \rightarrow \pi^*$ transition(s).[9] The $\pi \rightarrow \pi^*$ transitions are, by symmetry, all polarized in the plane of the bases. The current best estimates of the transition polarizations for guanine and cytosine are illustrated in Fig. 2.7.

An additional feature of nucleic acid structure that is particularly relevant for *LD* studies is the relative orientation of the bases to one another. Various sets of parameters are used to describe this. One such set for the base pairs of a duplex are illustrated in Fig. 3.1. Typical values for base roll, tilt, helical twist, and slide are given in Table 2.1. *LD* studies have shown that in solution significant variations from these standard values occur.

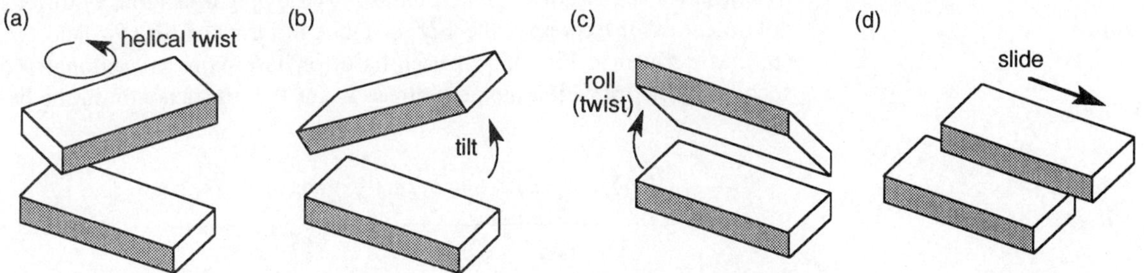

Fig. 3.1 (a) Helical twist, (b) tilt, (c) roll (alternatively called twist), and (d) slide of DNA base pairs with respect to one another.

DNA *LD*

From Table 2.1 it is apparent that the base rolls, tilts, and twists differ between the DNA polymorphs. DNA *LD* probes the orientation of base transition moments relative to the DNA helix axis (*i.e.* the roll and tilt). Thus, a standard B- or Z-DNA, where the base pairs are perpendicular to the helical axis of the DNA [$\alpha = 90°$ in eqn (3.1) for all transitions], would be expected to have *LD* signals for the base $\pi \to \pi^*$ of exactly the same spectral shape as the normal absorption (though of negative sign). The LD^r for such a DNA should therefore be constant across the base region of the spectrum (unless out-of-plane $n \to \pi^*$ transitions are present). As Fig. 3.4 shows, this is nearly true over the long wavelength band; however, it is not true for the shorter wavelength region of the spectrum. An effective value of 86° for α at 260 nm is often assumed for B-DNA based on the fibre diffraction structure and an estimate of contributions from different transitions to the total *LD*.

LD measures an average signal from all the bases of an oriented piece of DNA. The shortest piece of DNA that can be flow oriented is about 600 base pairs, though the orientation drops off sharply below about 1000 base pairs.

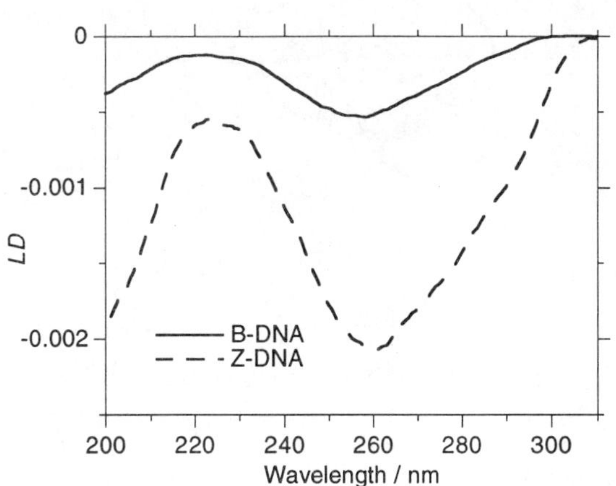

Fig. 3.2 *LD* spectra of poly[d(G-C)]$_2$ in 50 µM, 5 mM NaCl, 1 mM cacodylate. Z-form is induced by 50 µM [Co(NH$_3$]$^{3+}$.

In fact it now appears from recent flow *LD* experiments[10,11] that in solution the average tilt, even for B-DNA, is greater than this. By careful analysis of the different base transitions (assuming we know their transition polarizations) it has been possible to determine the average roll and tilt of the bases. The most commonly used equation is a variant of eqn (3.1) introduced by Matsuoka and Nordén[12] to account for overlapping transitions of different polarizations. It expresses the *LD^r* of DNA in terms of tilt (θ_X) and twist (θ_Y), as defined in Fig. 3.3, of each base (or base pair) and an angle (δ_i) specifying the transition moment direction for the *i*th transition in the base plane:

$$LD^r = \frac{3}{2} S \frac{\left\{ \sum_i F_i \varepsilon_i(\lambda) 3 \left[-\sin \delta_i \sin \theta_X + \cos \delta_i \cos \theta_X \sin \theta_Y \right]^2 - 1 \right\}}{\sum_i F_i \varepsilon_i(\lambda)} \tag{3.3}$$

where F_i is the fractional content of transition i (it depends on base composition), summation is taken over all transitions, and $\varepsilon_i(\lambda)$ is the extinction coefficient of transition i at wavelength λ.

Thus, normal B-DNA, which according to the classical fibre structure has $\theta_X = -2.1°$ and $\theta_Y = 4.0°$, should display only small variations in *LD^r* in the wavelength region 230–300 nm with $LD^r/S = -1.47 \pm 0.02$. A-form DNA, with $\theta_X = 19.3°$ and $\theta_Y = -3.2°$, is, however, expected to exhibit a sloping *LD^r*, with LD^r/S varying from about −1.2 at 230 nm to about −1.4 at 300 nm, somewhat depending on degree of exciton interactions.[12]

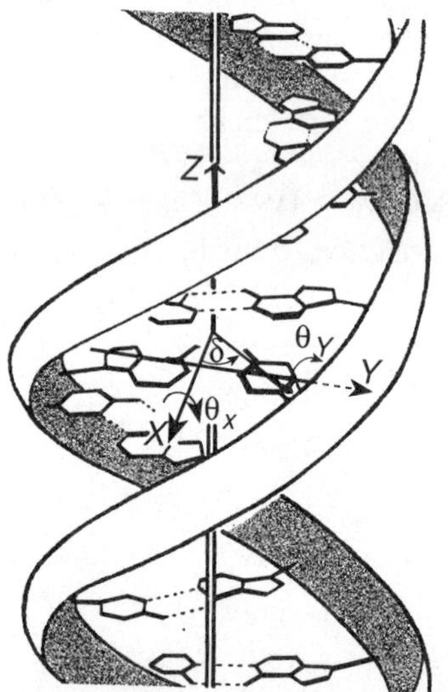

Fig. 3.3 Geometry for eqn (3.3).

From knowledge of the band shapes of absorption components [$\varepsilon_i(\lambda)$] for each transition and the polarization directions (δ_i) in each base it has also been possible in a few cases to fit the experimental *LD^r* spectrum with one calculated according to eqn (3.3) using trial values of tilt, twist and orientation factor. For example, the strongly wavelength-dependent *LD^r*

observed for DNA in presence of silver(I) ions has been reproduced using a DNA conformation with extreme base inclinations away from perpendicularity: $\theta_X = 47°$ and $\theta_Y = -24°$.[13]

In any analysis of DNA *LD* the most difficult part is often determining the degree of orientation of the DNA in a given experiment. The theoretical limiting value of an LD^r signal of a transition polarized perpendicular to an orientation axis is -1.5. However, this value is never reached with DNA. Calf thymus DNA in a cylindrical coaxial flow device has a theoretical maximum LD^r value of -1.48 at 260 nm,[12] however, values of -0.2 in water at shear gradients of 3000 s^{-1} are typical. In principle, at least, S may be calculated. In practice, for macromolecular systems, S is usually determined more or less reliably by empirical means either by calibrating with known maximum LD^r values, or by calibrating with a probe dye whose binding geometry is known.[14] With flow orientation, laminar flow must be retained to avoid scattering effects.[15]

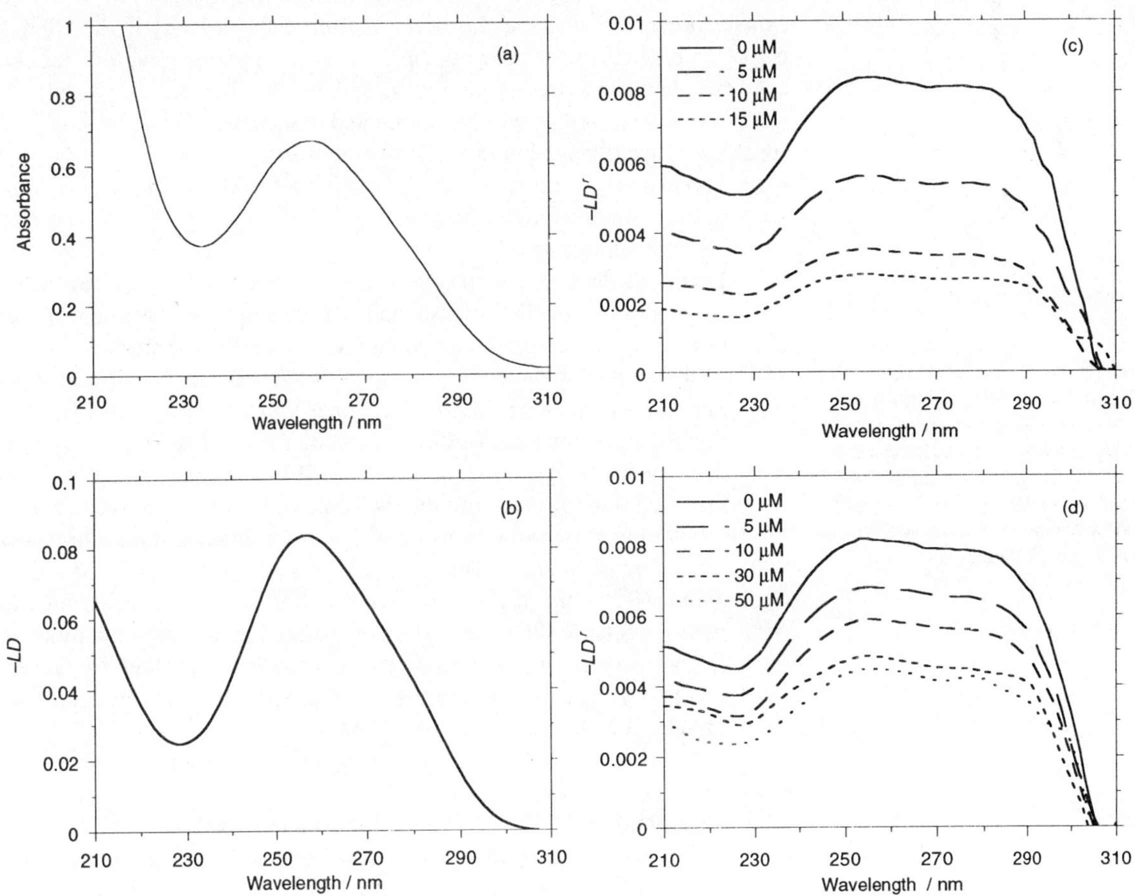

Fig. 3.4 Spectra of calf thymus DNA (100 µM in 5 mM NaCl) with spermine (concentrations indicated in figure). (a) Normal absorption; (b) *−LD*, (c) *−LD^r*. (d) shows the corresponding *−LD^r* spectra in the presence of $[Co(NH_3)_6]^{3+}$.

The magnitude of S can be used to provide information about the length, or flexibility, or bendability of a piece of DNA. If S increases as the result of some change imposed on the system (such as the addition of a DNA binding ligand), then the DNA has become more oriented. This means it is either lengthened, or stiffened, or some bend or kink has been removed. More information about the system is required to distinguish these effects. Fig. 3.4 shows the effect of the gradual addition of some amines to DNA. Spermine is a molecule that is known to bend DNA and is an important factor in the condensation of DNA into chromosomes.

LD of DNA-bound ligands

In most situations not all the ligands will be bound to DNA, so the isotropic absorbance is a mixture of that due to bound and free ligands. Thus care must be taken in calculating LD' to remove the absorbance due to unbound ligands.

Once the orientation of DNA itself is understood a next step is to use *LD* to probe the binding and orientation on DNA of small ligand molecules. A key feature of such *LD* studies is that if there is no specific binding, then there will be no *LD* of the ligand transitions. Conversely, the presence of an *LD* signal from an absorption band of the putative ligand immediately implies that it is bound. Although *LD* cannot tell us *where* a ligand binds to DNA, it can be used to determine the *orientation* of the ligand (if the ligand's transition moments have been assigned). This often provides significant clues as to the binding mode. A ligand may bind to DNA in a number of different types of sites, as well as with sequence and orientation preferences. Broadly speaking a ligand may bind in the following ways.

- **Externally** bound to the phosphate backbone: this is usually orientationally fairly non-specific and the induced *LD* (and *CD*) is correspondingly small.
- **Intercalated** between DNA bases: this mode requires planar molecules and the DNA to unwind and open up a slot between adjacent base pairs so the ligand may be sandwiched between them.

Minor groove binders tend to be selective for A-T rich sequences as the electrostatic potential of the minor groove is deeper for these sequences than for G-C rich ones and also guanine sterically hinders this binding mode as its amine group pokes out into the groove (Fig. 2.6b).

- In the **minor groove**: this mode is frequently adopted by aromatic molecules containing internuclear bonds with some rotational freedom. For example, long molecules such as netropsin, 9-hydroxyellipticene, distamycin, Hoechst 33258, and DAPI (Fig. 3.5), which can fit snugly into the minor groove with the molecule following the curvature of the groove but are too large to fit well into an intercalation site, favour this mode of binding. The long axis of such molecules is oriented at about 45° to the helix axis and the short axis polarized transition perpendicular to it (so approximately parallel to the base transitions).
- In the **major groove**: this type of binding is found for several regulatory proteins and there is enough space to accommodate most smaller ligands in a variety of orientations.

The application of *LD* for studying DNA–ligand systems is best seen by considering some examples. We shall first examine a classical intercalator: ethidium bromide (EB, Fig. 3.5). Ethidium is a monovalent cation known to be bound by intercalation to various sequences of double stranded B-form DNA. Two of its absorption bands are accessible outside the absorption region of DNA: one due to a transition occurring at 285–300 nm with a polarization approximately parallel to the long axis of the phenanthridinium moiety and one at 450–500 nm with a polarization (still in the plane of the molecule) at an angle 60° from the long axis. Fig. 3.6 shows the isotropic absorption spectrum and the reduced linear dichroism spectra of a mixture of calf thymus DNA and EB. The following LD^r values can be noted for the various chromophores and transitions:

Wavelength / nm	250–290	330	500
Transition identity	DNA	EB, long axis	EB, 60° polarization
LD^r	−0.23	−0.22	−0.31

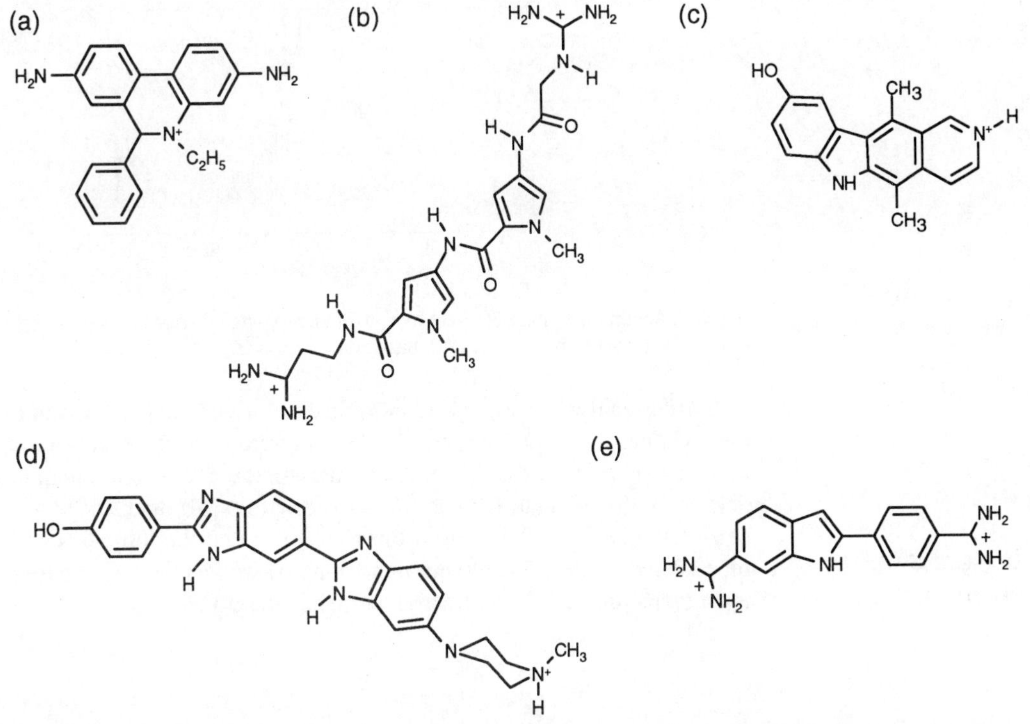

Fig. 3.5 (a) The intercalator ethidium bromide, and (b)–(e) some minor groove binders: (b) netropsin, (c) 9-hydroxyellipticene, (d) Hoechst 33258, and (e) 4',6-diamidino-2-phenylindole (DAPI). The smaller molecules DAPI and 9-hydroxyellipticene may also intercalate.

A first qualitative conclusion is that the strongly negative LD^r values for both of the in-plane polarized transitions of EB are consistent with the plane of the latter being perpendicular to the helix axis of DNA. Secondly, the similar LD^r magnitude in the nucleobase absorption region supports the conclusion that the bases and EB are coplanar. (That the LD^r is constant in the 250–290 nm region is consistent with the B conformation of DNA as discussed above.) Similar magnitude LD^r signals for the 250–290 nm DNA and the 330 nm EB transitions suggest that the EB long axis lies along the average polarization direction for these transitions, whereas the larger value for the 60° polarized EB transition suggests it lies somewhat more perpendicular to the average DNA helix axis.

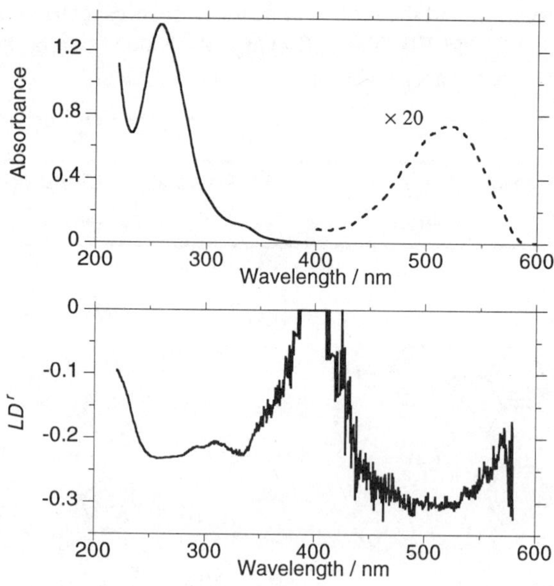

Fig. 3.6 Absorbance and LD^r spectra of ethidium bromide in 5 mM phosphate buffer, pH = 6.9, bound to calf thymus DNA , DNA base : drug ratio is 20 : 1.[16]

Further analysis can provide more geometric information; assuming that the most negative LD^r value (of the 330 nm transition) represents a transition that is oriented perfectly perpendicular to the helix axis enables us to determine the orientation factor S. Insertion of $\alpha = 90°$ and $LD^r = -0.31$ into eqn (3.1) gives $S = 0.21$. From this it follows that the long axis of EB (as represented by the 250–290 nm transition), as well as the nucleobase planes, must be inclined at about 20° from perpendicular to the helix axis.

Inserting $LD^r = -0.22$, $S = 0.21$ into eqn (3.1) gives $\alpha = 71°$.

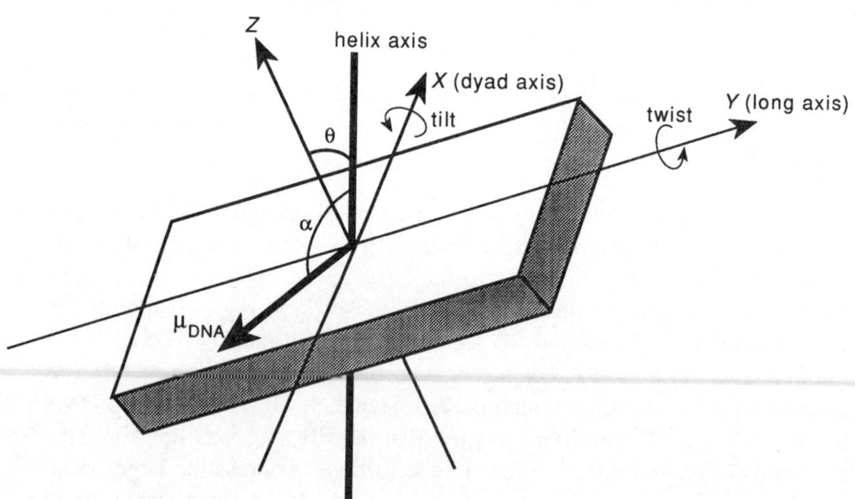

Fig. 3.7 Average orientation of DNA bases with respect to the helix axis.

It has been observed for other intercalating dyes, such as 9-aminoacridine and quinacrine, that their long axis, which is also believed to be oriented parallel to the longest dimension of the base-pair pocket, is inclined some 20° from perpendicularity, whereas their short axis is oriented perpendicular to the helix axis. It may thus be inferred that the long axis of the base pairs is rotated by 20° about the short axis, the latter coinciding with the pseudo-dyad axis of DNA (Fig. 3.7).

Our next example refers to another monocationic dye molecule DAPI (Fig. 3.5), which is known to prefer AT sequences of duplex DNA into which it binds in the minor groove. A strong absorption band near 360 nm is known to be polarized near the long axis of this dye. Recording the flow linear dichroism in the DNA region as well as for the dye gives the following values of LD^r at base : dye ratios of approximately 50 : 1 in mixtures with calf-thymus DNA and the alternating duplexes poly[d(A-T)]$_2$ and poly[d(G-C)]$_2$.

DNA	calf thymus DNA	poly[d(A-T)]$_2$	poly[d(G-C)]$_2$
LD^r {Wavelength / nm}	−0.18 {260}	−0.028 {360}	−0.033 {260}
	+0.09 {360}	+0.012 {360}	−0.030 {370}

A number of conclusions may be drawn for this data that illustrate the various pieces of information that may be provided by flow LD studies of drug–DNA systems.[11]

- Assuming that the DNA bases are perpendicular to the helix axis, *i.e.* $\alpha = 90°$, allows us to gauge the orientation parameter S in each of these experiments: $S = 0.12$, 0.019 and 0.022 for the three biopolymer samples, respectively. The markedly lower degree of alignment by the flow for the synthetic polynucleotides is a result of their significantly smaller contour lengths.

  Using more accurate values of DNA base α such as 86° (B-form fibre structure) or the recently inferred tilt for the 260 nm transitions corresponding to $\alpha = 71°$ will not make a large difference to the final geometrical conclusions.

- Using these S values we may now characterize the binding geometries of the DNA–drug complexes in terms of the angle between the long axis of the drug molecule and the DNA helix axis. Inserting the LD^r values observed for the long axis polarized band of DAPI around 360–370 nm we get:

calf-thymus DNA–DAPI	$\alpha = 45°$
poly[d(A-T)]$_2$–DAPI	$\alpha = 46°$
poly[d(G-C)]$_2$–DAPI	$\alpha = 80°$

As mentioned earlier, a 45° orientation for the long axis of a drug is expected for a minor groove binding geometry so we conclude that this is indeed the binding mode of DAPI in alternating poly[d(A-T)]$_2$ as well as in mixed sequence DNA at these binding ratios. In alternating poly[d(G-C)]$_2$, however, as a consequence of the exocyclic amino group of guanine protruding into the minor groove, a different binding mode is found. DAPI here is oriented almost perpendicular to the helix axis and also shows a larger hypochromicity and stronger red-shift of the ligand transition than with the other two DNAs. These observations are suggestive of an intercalative binding mode, despite the size of the DAPI molecule.

From the similar orientation and red-shift of DAPI in presence of calf thymus DNA and poly[d(A-T)]$_2$ we may conclude that for the rather low binding ratios at which the above data were determined, DAPI is bound to AT stretches of the mixed sequence DNA. In fact, at higher binding ratios a negative contribution to the *LD* is evidence for some binding in the poly[d(G-C)]$_2$ binding mode.[17]

This example thus illustrates the potency of *LD* to discriminate between groove binding and intercalation and, in addition, demonstrates that a given DNA ligand may bind in several different ways depending on nucleobase sequence and binding ratio. *LD*, as *CD*, may also be used to directly determine equilibrium binding constants for DNA–drug systems. Some methods to do this are outlined in Appendix 2.

References

(1)   Rodger, A. *Meth. Enzymol.* **1993**, *226*, 232

(2)   Griebenow, K.; Holzwarth, A. R.; van Mourik, F.; van Grondelle, R. *Biochim. Biophys. Acta* **1991**, *1058*, 194

(3)   van Amerongen, H.; van Grondelle, R. *J. Mol. Biol.* **1989**, *209*, 433

(4)   Wan, C.; Qian, J.; Johnson, C. K. *Biochemistry* **1991**, *30*, 394

(5)   Nordén, B.; Lindblom, G.; Jonás, I. *J. Phys. Chem.* **1977**, *81*, 2086

(6)   Kubista, M.; Hagmar, P.; Nielsen, P. E.; Nordén, B. *J. Biomol. Struct. Dyn.* **1990**, *8*, 37; Hagmar, P.; Marquet, R.; Colson, P.; Nielsen, P.; Nordén, B.; Houssier, C. *J. Biomol. Struct. Dyn.* **1989**, *7*, 19

(7)   Hagmar, P.; Nordén, B.; Baty, D.; Chartier, M.; Takahashi, M. *J. Mol. Biol.* **1992**, *226*, 1193

(8)   Takahashi, M.; Kubista , M.; Nordén, B. *Biochimie* **1991**, *73*, 219

(9)   Matsuoka, Y.; Nordén, B.; *Biopolymers* **1982**, *21*, 1731; Lyng, R.; Rodger, A.; Nordén, B. *Biopolymers* **1992**, *31*, 1709; *32*, 1201

(10)   Chou, P.-J.; Johnson, W. C. Jr. *J. Amer. Chem. Soc.* **1993**, *115*, 1205

(11)   Nordén, B.; Kubista, M.; Kuruscev, T. *Q. Rev. Biophysics* **1992**, *25*, 51

(12)   Matsuoka, Y.; Nordén, B. *Biopolymers* **1982**, *21*, 2433

(13)   Nordén, B.; Matsuoka, Y.; Kurucsev, T. *Biopolymers* **1986**, *25*, 1531

(14)   Hiort, C.; Nordén, B.; Rodger, A. *J. Amer. Chem. Soc.* **1990**, *112*, 1971

(15)   Lee, C. S.; Davidson, N. *Biopolymers* **1968**, *6*, 531

(16)   Tuite, E.; Nordén, B. *Biorg. Med. Chem.* **1995**, *3*, 701

(17)   Eriksson, S.; Kim, S. K.; Kubista, M.; Nordén, B. *Biochemistry,* **1993**, *32*, 2987

4 Linear dichroism of small molecules

4.1 Introduction

LD spectra are quickly measured and usually easy to interpret. However, not many research groups use *LD* because the equipment required for orienting samples is not currently commercially available. An aim in this book is to show that establishing *LD* facilities is fairly straightforward and that the effort is well recompensed by the data that can be collected.

In the previous chapter we focused on applications of *LD* to biological macromolecular systems. When *LD* is used to study small molecules, a much more detailed analysis of the data is often possible. The resulting structural and spectroscopic information may be of direct and immediate use or may be the input for analysis of other experiments such as the *LD* of the small molecule bound to DNA.

In this chapter we shall see how to analyse an *LD* spectrum. As usual in science, we shall introduce simplifying assumptions and, depending on what we are after and what information we have, shall end up with a number of special cases, each appropriate to a specific class of applications. References 1–9 will enable much of the *LD* literature to be discovered.

Most published *LD* data are for electronic transitions using visible or ultraviolet light, however, the analysis is exactly the same for vibrational transitions.

4.2 Orientation distribution

In §1.4 a variety of different orientation methods were mentioned. In practice, with the possible exception of a crystalline environment, the orientation is never perfect and we need to use the concept of an orientation distribution. In the previous chapter the averaging inherent in the methods used for macromolecules meant that we could assume local uniaxial orientation and use the simple orientation parameter S. More generally this is not the case.

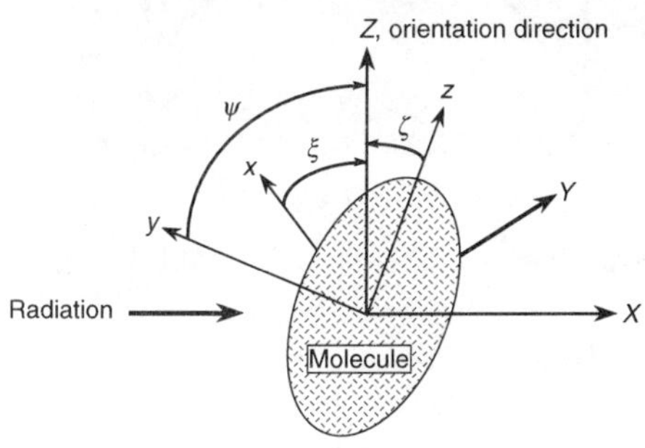

Fig. 4.1 Macroscopic and molecular axis systems and angles relating them.

Note we use the small letters $\{x,y,z\}$ for the internal molecular axis system and the capital letters $\{X,Y,Z\}$ for the macoscropic axis system.

We first define a laboratory (macroscopic) axis system, $\{X,Y,Z\}$. $\{X,Y,Z\}$ is defined by the experiment: we take X to be the direction of propagation of the light and Z to be the macroscopic direction of orientation. Y then completes the right-handed axis system.

When studying *LD* it is also convenient to define the molecular axis system, $\{x,y,z\}$, using the *LD* experiment. We take z to be the molecular orientation axis. This is the axis in the molecule which has maximum value of

$$\left\langle \cos^2 \zeta \right\rangle \tag{4.1}$$

where ζ (zeta) is the angle between z and Z and $\langle\ \rangle$ denotes average over all the molecules in the sample. x is the axis perpendicular to z that has the smallest value of

$$\left\langle \cos^2 \xi \right\rangle \tag{4.2}$$

where ξ (xi) is the angle between x and Z. ψ (psi) is similarly the angle between y and Z. In most chemical applications the molecular axis system is defined using symmetry, with z being the highest order symmetry-determined axis. For many molecules this definition and the one we use for *LD* do not coincide. The two *LD* axis systems are illustrated in Fig. 4.1.

Although few *LD* experiments involve perfect orientation, the orientation distributions generally have a degree of symmetry due both to the symmetry properties of the external orienting force and also to the structural symmetry of the molecules themselves. Consider first the external orienting force. Many *LD* experiments have *uniaxial orientation* which means that the probability of finding a given molecular axis at an angle θ from the macroscopic orientation direction, Z, is constant around a cone centred about Z (Fig. 4.2). For some experiments, however, including rectangular squeezing of a gel (Fig. 1.9) and when the light is incident radially upon a cylindrical couette flow cell (Fig. 1.8) the system behaves as if the orientation were *biaxial* (Fig. 4.2) for which the constant probability contour for a given molecular axis traces out an ellipse.

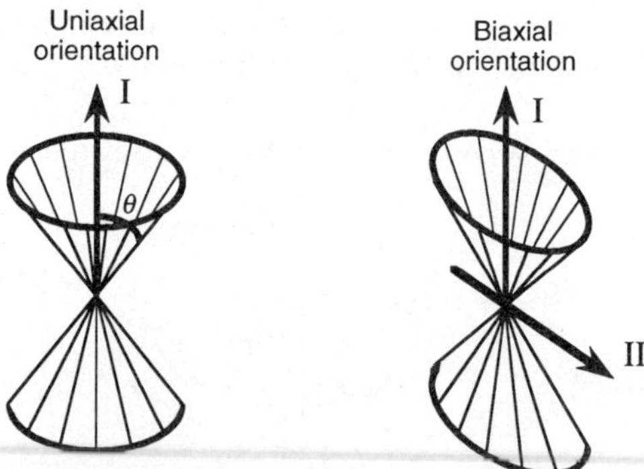

Fig. 4.2 Uniaxial and biaxial orientation.

How the external orienting force operates on the molecule depends on the nature of the molecule. In some experiments it is easy to determine the molecular orientation axis, z. Under electric field orientation, z is the direction of the permanent or induced dipole moment (Fig. 4.3). Orientation methods based on steric or shear forces usually have z corresponding with the axis of preferred alignment, *i.e.* the 'longest' axis. However, this is not necessarily so. Imagine a planar, rectangular molecule. The orientational distribution function for this may be bimodal with the two diagonals having equal, maximum, tendency to be parallel to the orientation direction, but the axis with maximum $\langle \cos^2 \zeta \rangle$ is the one bisecting the two diagonals; hence, the orientation axis is parallel to the long sides of the rectangle (Fig. 4.3). Such situations usually correspond with molecules where symmetry defines the axes (see §4.5). In the case of flow-oriented DNA, although the orientation method is biaxial, the helical symmetry of the molecule renders the average local orientation to be uniaxial. For low symmetry molecules in anisotropic hosts such as a liquid crystal or a stretched polymer matrix, the identity of z may be far from clear and can only be determined by experiment.

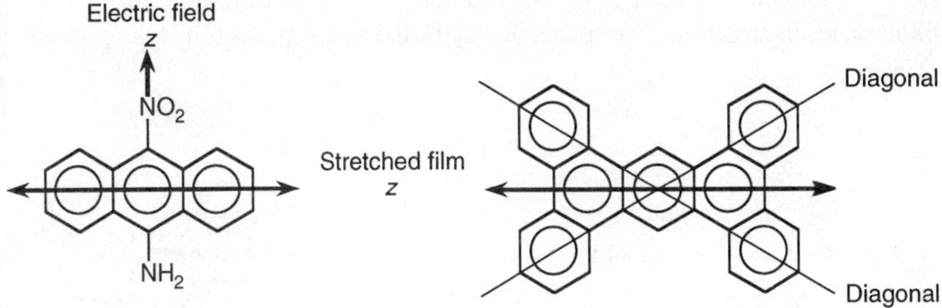

Fig. 4.3 Molecular orientation axes, z, under electric field orientation and stretched film orientation.

4.3 Some *LD* definitions

General definition of LD

In eqn (1.2) we defined *LD* as the difference in absorption of two linearly polarized light beams that are at right-angles to one another and to their direction of propagation:

$$LD = A_{\parallel} - A_{\perp}$$

This notation is normally reserved for uniaxial samples with $\parallel$ denoting the direction of the unique axis. So, although most *LD* experiments have effective uniaxial orientation, in this chapter we shall use the more general definition

$$LD = A_Z - A_Y \tag{4.3}$$

where A_Z is the absorbance of Z-polarized light, and similarly A_Y. In §1.2 we defined the *transition moment* to be a vectorial property (see Appendix 1) having a well-defined direction (the transition polarization) within each molecule and a well-defined length (which is proportional to the square root of the absorbance). We denote the transition moment by the vector $\boldsymbol{\mu}$. A_Z is then proportional to the square of the Z-component of $\boldsymbol{\mu}$, so

We define X to be the propagation direction of the electromagnetic radiation throughout this book.

$$LD = K\left(\left\langle \mu_Z^2 \right\rangle - \left\langle \mu_Y^2 \right\rangle\right) \tag{4.4}$$

where $\langle \; \rangle$ denotes an average over the orientational distribution of the molecules in the sample and K is constant.

Reduced linear dichroism

The reduced dichroism, LD^r as defined in eqn (3.1) may then be written

$$LD^r = 3\left(\frac{A_Z - A_Y}{A_X + A_Y + A_Z}\right) = 3\left(\left\langle \hat{\mu}_Z^2 \right\rangle - \left\langle \hat{\mu}_Y^2 \right\rangle\right) \tag{4.5}$$

The factor of 3 in eqn (4.5) arises because the isotropic absorbance

$$A = \frac{1}{3}\left(A_X + A_Y + A_Z\right)$$

is a rotational average in three dimensions.

where $\hat{\mu}_Z$ is the Z-component of the unit vector along $\boldsymbol{\mu}$, $\hat{\boldsymbol{\mu}}$. $\hat{\mu}_Z$ is thus the cosine of the angle between the transition moment polarization and the Z-axis, similarly $\hat{\mu}_Y$ is the cosine of the angle between $\boldsymbol{\mu}$ and the Y-axis. The advantage of LD^r over LD in analysis is that dependence both on the concentration of the sample and the dipole strength, μ^2, of the transition has been removed. Also, across the absorption envelope of *a single transition* the LD^r should be constant (Fig. 4.4). Thus, LD^r is a function only of the geometric arrangement in space of the transition moments relative to the orientation axis.

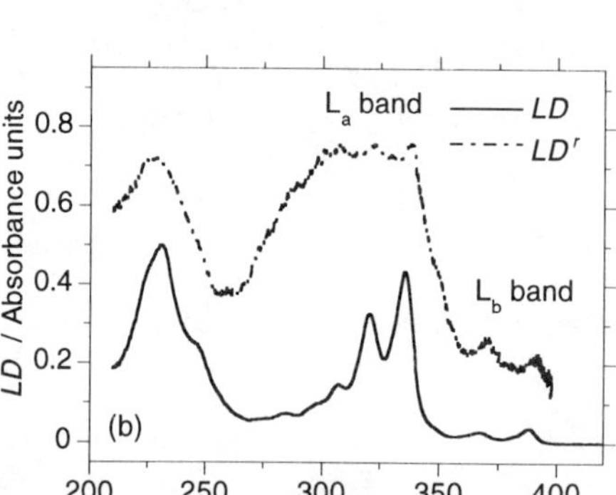

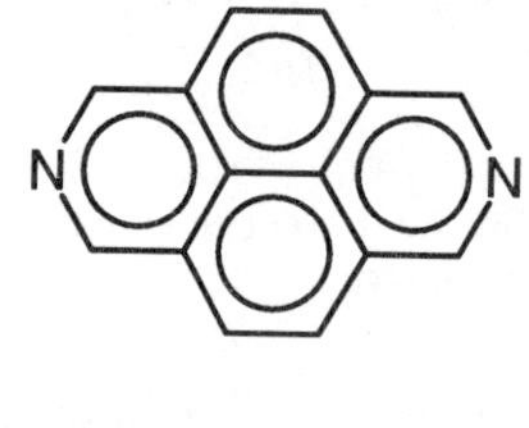

Fig. 4.4 *LD* and *LD*^{r} of 2,7-diazapyrene oriented in a stretched (5 ×) PVA film. The constant *LD*^{r} signal from 300–350 nm shows that the L$_a$ transition has a single polarization.[1]

We could finish this chapter here since in principle eqn (4.5) is the only relation we need for interpreting an *LD* experiment. To use it we would first propose a model for how the chromophoric groups are arranged in space and how the transition moments are directed within each chromophore, then calculate the LD^r using eqn (4.5) and finally average over all chromophores and all transitions. If the calculated LD^r spectrum did not agree with the experimental one, then the model would have to be revised. With many systems, by exploiting different parts of the spectrum and differently polarized transitions, our confidence in the structural conclusions from this approach would be high and it is in fact the preferred procedure for an object whose macroscopic *and* microscopic structure is well defined, such as a crystal. However, a more common situation is one in which there is a superimposed uncertainty due to poor macroscopic alignment of the main structural entities composing the sample. For example, in a polymer chain where our major concern is the structure of each link we may only have information about the global orientation of the polymer. For many applications of the latter type it is convenient to factorize the reduced dichroism into a product between an *optical factor*, O, which depends on how the transition moment is oriented in a local coordinate system and an *orientation factor*, S, which depends on how well the molecule is macroscopically oriented.

The orientation factor

The orientation factor, S, is generally defined in such a way that a sample oriented perfectly parallel with the macroscopic orientation axis of the experiment will have $S = 1$, and a randomly oriented sample will have $S = 0$. In the previous chapter we simply used it as a scaling factor; it can, however, be used to provide more information about the system.

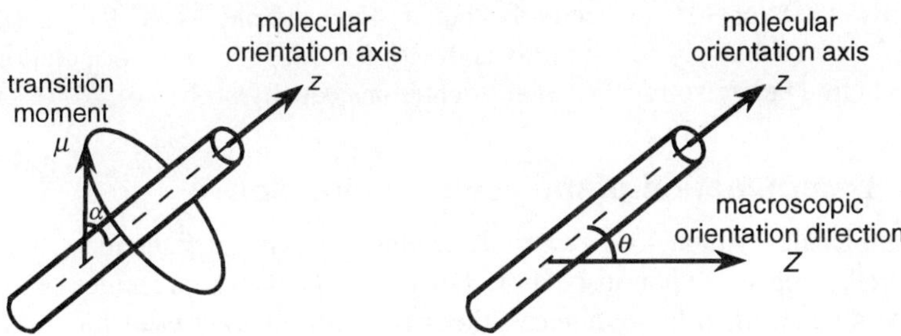

Fig. 4.5 Geometry used for uniaxial orientation of rod-like molecules.

When the orientation method used in an LD experiment provides a uniaxial sample *and* the sample molecules are also characterized by a unique axis around which all orientations are equally probable (so the molecule is 'rod-like'), we may then write:

$$S = \tfrac{1}{2}\left(3\langle\cos^2\theta\rangle - 1\right) \tag{4.6}$$

where θ is the angle between the macroscopic orientation direction Z and the molecular orientation axis z. We use θ instead of ζ—to which we gave the same definition in §4.2—to remind us that eqn (4.6) only holds for rod-like molecules in uniaxial samples. Further, if the transition moment lies at an angle α (see margin) relative to z we have the following expression for the LD^r:

$$LD^r = 3\left\{\tfrac{1}{2}\left(3\cos^2\alpha - 1\right)\right\} \times \left\{\tfrac{1}{2}\left(3\langle\cos^2\theta\rangle - 1\right)\right\} \tag{4.7}$$

$$= 3O(\alpha)S(\theta)$$

Eqn (4.7) is derived in the next section.

In defining α we do not need to specify upon which side of z it lies, however, we do need to decide this before assigning a transition moment polarization. We return to this in §4.5.

In principle, S may be used to examine the mechanism of orientation of the sample once the optical factor (*i.e.* the transition moment direction α) is known. Conversely, if S can be assessed independently, the transition moment direction in the molecular frame may be determined. These two possibilities illustrate the two major classes of applications of LD

spectroscopy: the structural and the spectroscopic applications which form the subject matter, respectively, of §4.4 and §4.5.

Effective angle

How to interpret averages such as $\langle \cos^2 \theta \rangle$ in terms of structure is far from clear. If we know α, we can determine S from the LD^r using eqn (4.7). Unfortunately, this does not mean that we thereby have an exact value for the angle θ at which the molecules may be oriented. First of all the cosine square is associated with a sign ambiguity (we do not know whether it is $+\cos \theta$ or $-\cos \theta$) and secondly, the distribution of θ values, over which the $\cos^2$ function is averaged, may have features that are poorly characterized by the single angle value, θ_{eff} (denoting θ effective), defined by:

$$\cos^2 \theta_{eff} = \langle \cos^2 \theta \rangle \tag{4.8}$$

For $\langle \cos^2 \theta \rangle$ values close to 1 or 0 we can with some confidence say that θ is close to $0°$ or $90°$, respectively, but if, for example, $\langle \cos^2 \theta \rangle = \frac{1}{3}$ (*i.e.* $S = 0$) we cannot say whether this is due to $\theta = 54.7°$ or to the orientational distribution being isotropic (*i.e.* all orientations equally probable).

4.4 Determination of molecular orientation

For molecules with an axis system determined by symmetry the identity of z is usually apparent upon inspection. For lower symmetry molecules (one or no axis determined by symmetry) the orientation method used for the LD *always* defines a unique z, but we cannot easily identify z upon inspection. In this section we shall first give the general equations, then show how to identify $\{x, y, z\}$ for the particular example of riboflavin.

In solving any particular problem it is usually convenient first to choose a temporary molecular axis system $\{x', y', z'\}$ which probably bears no relationship to the orientation axis system. Analogously to the definitions we used with the 'true' molecular axis system, we may express the orientation of the macroscopic orientation axis Z by the vector

$$\left(\cos \xi', \cos \psi', \cos \zeta'\right)_{x'y'z'} \tag{4.9}$$

in the $\{x', y', z'\}$ axis system.

For molecular orientation in a macroscopically anisotropic sample that has uniaxial orientation we shall also find it convenient to define orientation tensors (or matrices), S and S', for respectively the true and the temporary axis systems.

$$S = \begin{pmatrix} S_{xx} & S_{xy} & S_{xz} \\ S_{yx} & S_{yy} & S_{yz} \\ S_{zx} & S_{zy} & S_{zz} \end{pmatrix}$$

$$= \begin{pmatrix} \frac{1}{2}\left(3\langle \cos^2 \xi \rangle - 1\right) & 0 & 0 \\ 0 & \frac{1}{2}\left(3\langle \cos^2 \psi \rangle - 1\right) & 0 \\ 0 & 0 & \frac{1}{2}\left(3\langle \cos^2 \zeta \rangle - 1\right) \end{pmatrix} \tag{4.10}$$

Margin notes:

54.7° is often called the 'magic angle' since in many experimental situations, including *LD* and solid state NMR, orientation at this angle leads to many of the features of a full rotational average.

The true molecular orientation axis system $\{x,y,z\}$ in a uniaxial system is the one which has a diagonal orientation tensor, **S**. The label *diagonal axis* or *diagonal coordinate system* is often used for $\{x,y,z\}$ because its **S** tensor is diagonal.

Since we defined z to be the most oriented molecular axis, and x to be the least oriented it follows that:

$$S_{zz} \geq S_{yy} \geq S_{xx} \tag{4.11}$$

Further, the trace of the orientation tensor is zero so:

$$S_{xx} + S_{yy} + S_{zz} = 0 \tag{4.12}$$

For the temporary axis system we have:

$$S' = \begin{pmatrix} S_{x'x'} & S_{x'y'} & S_{x'z'} \\ S_{y'x'} & S_{y'y'} & S_{y'z'} \\ S_{z'x'} & S_{z'y'} & S_{z'z'} \end{pmatrix}$$

$$= \begin{pmatrix} \frac{1}{2}\left(3\langle\cos^2\xi'\rangle - 1\right) & 3\langle\cos\xi'\cos\psi'\rangle & 3\langle\cos\xi'\cos\zeta'\rangle \\ 3\langle\cos\xi'\cos\psi'\rangle & \frac{1}{2}\left(3\langle\cos^2\psi'\rangle - 1\right) & 3\langle\cos\psi'\cos\zeta'\rangle \\ 3\langle\cos\xi'\cos\zeta'\rangle & 3\langle\cos\psi'\cos\zeta'\rangle & \frac{1}{2}\left(3\langle\cos^2\zeta'\rangle - 1\right) \end{pmatrix} \tag{4.13}$$

We could now write the general equations for determining $\{x,y,z\}$ from $\{x',y',z'\}$. In practice we usually only need to consider planar molecules.

Uniaxially oriented planar aromatic molecules with only in-plane polarized transitions

Many chromophores of interest to chemists and biochemists are molecules whose only symmetry property is a reflection plane which contains all the atoms. When such a molecule is uniaxially oriented, as in a stretched film or a nematic liquid crystal, although it is in principle possible that the molecule is oriented so that its reflection plane is perpendicular to the macroscopic orientation direction, it is generally true that the orienting forces align the molecules so that they present a minimum cross-sectional area to the orienting force, *i.e.* to Z. Thus x is perpendicular to the molecular plane and y and z are somewhere within the plane of the molecule, but we do not know where (though z is usually close to the 'longest' axis of the molecule) (Fig. 4.6). What follows is how we can determine z if we have a number of transitions whose polarizations in the molecular plane we know. Although our emphasis in this book is on electronic transitions, we note that infra-red *LD* is particularly useful here since polarizations of some bond-stretch transitions are obvious.

Because we do not know where z lies within the molecule, as discussed above, we start by choosing a convenient temporary axis system $\{x',y',z'\}$ and immediately note that $x' = x$ so $\xi' = \xi$ for a planar molecule in a uniaxially oriented system. The LD^r expressed in terms of parameters in the $\{x',y',z'\}$ axis system is then

$$LD^r = 3\left(S_{y'y'}\sin^2\alpha' + S_{z'z'}\cos^2\alpha' + S_{y'z'}\sin\alpha'\cos\alpha'\right) \tag{4.14}$$

(see below for derivation) where α' is the angle between the transition dipole moment and axis z'. The three orientation parameters are three unknowns in this equation. If our temporary molecular axes actually proved to be the same as the orientation-determined molecular axes, then $S_{y'z'} = 0$. Otherwise we need three transitions to determine the unknowns and hence the molecular orientation axis. Once we have done that we use

$$\tan(2\beta) = \frac{S_{y'z'}}{S_{z'z'} - S_{y'y'}}$$

(4.15)

(see below for derivation) to determine β, the angle between z' and z (Fig. 4.6). The true molecular orientation axis system $\{x, y, z\}$ is then identified.

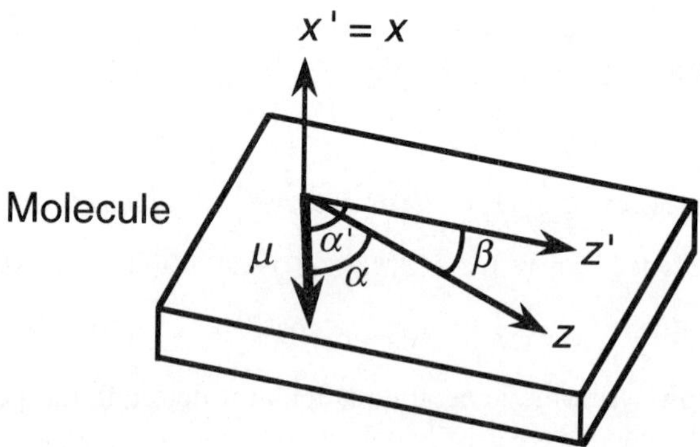

Fig. 4.6 Orientation of a planar molecule in a stretched film.

The equation analogous to eqn (4.14) for the true axis system, $\{x, y, z\}$, for in-plane polarized transitions of a planar molecule in a uniaxial sample is then

$$LD^r = 3\left(S_{zz} \cos^2 \alpha + S_{yy} \sin^2 \alpha\right)$$

(4.16)

As a special case of eqn (4.16) we note that a rod-like molecule in a uniaxial sample would have

$$S_{xx} = S_{yy} = -\frac{1}{2} S_{zz}$$

(4.17)

Eqn (4.7) then follows from eqn (4.16) if we write $S_{zz} = S(\theta)$.

Uniaxial orientation of any molecule

In the previous subsection we restricted our attention to planar molecules with no out-of-plane transitions. For the more general case of uniaxial orientation of any molecule the following equation holds in the diagonal (*i.e.* true) coordinate system:[2]

$$LD^r(\lambda) = \frac{LD(\lambda)}{A(\lambda)} = 3\left(\frac{S_{zz}\varepsilon_z(\lambda) + S_{yy}\varepsilon_y(\lambda) + S_{xx}\varepsilon_x(\lambda)}{\varepsilon_z(\lambda) + \varepsilon_y(\lambda) + \varepsilon_x(\lambda)}\right)$$

(4.18)

where A is the absorbance of an equivalent (same pathlength and same sample concentration) unoriented sample and $\varepsilon_z(\lambda)$ *etc.* are the extinction coefficients (at wavelength λ) corresponding to light polarized parallel to the respective coordinate axis. We shall return to eqn (4.18) later as it is particularly useful for analysing the LD spectra of small symmetric molecules, but emphasize that it is quite general holding for complicated, low-symmetry molecules with overlapping transitions.

Determination of the molecular orientation axis of riboflavin in a stretched film

Riboflavin has three infra-red bond stretching vibrations whose polarizations can readily be determined. As illustrated in Fig. 4.7 we choose our temporary axis system so that one of the carbonyl stretching transition moments has $\alpha' = 0$. The values of the LD^r for each transition are also given in the figure.

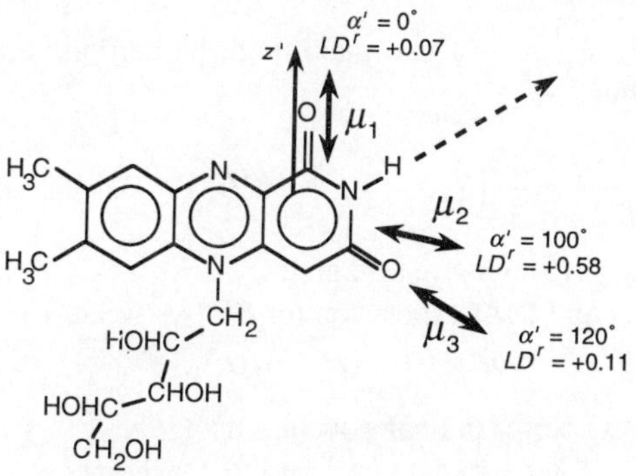

Fig. 4.7 Riboflavin indicating the temporary axis system and the polarizations and LD^r values for the three transitions used in the text. Data are from reference 10.

We begin our calculations by substituting the experimental data into eqn (4.14) for each transition. Note, the orientation parameters are molecular properties.

$$+0.070 = 3\left(S_{z'z'}\right)$$

$$+0.580 = 3\left(0.030 S_{z'z'} + 0.970 S_{y'y'} - 0.174 S_{y'z'}\right) \tag{4.19}$$

$$+0.110 = 3\left(0.250 S_{z'z'} + 0.750 S_{y'y'} - 0.433 S_{y'z'}\right)$$

From which it follows that

$$S_{z'z'} = 0.024$$
$$S_{y'y'} = 0.256 \tag{4.20}$$
$$S_{y'z'} = 0.373$$

Substituting the orientation parameters into eqn (4.15) tells us that β either equals 61° or 151°. By inspection of Fig. 4.7, noting the orientation influence of the substituent chain at the bottom, we deduce that the value of 61° is the correct one and the dashed line of Fig. 4.7 is the molecular orientation axis z. Once the orientation has been determined, the polarizations in the true axis system of electronic (and other vibrational) transitions can be calculated using eqn (4.16).

Derivation of eqns (4.14) and (4.15) for determining the molecular orientation axis of a planar molecule in a uniaxially oriented system

In the temporary molecular axis sytem, $\{x', y', z'\}$, the molecular orientation axis, z, may be written as the vector (Fig. 4.6)

$$(0, \sin\beta, \cos\beta)_{x'y'z'} \tag{4.21}$$

Thus, from eqn (4.9) and the definition of ζ it follows that

$$\cos\zeta = \mathbf{Z} \cdot \mathbf{z} = \cos\psi' \sin\beta + \cos\zeta' \cos\beta \tag{4.22}$$

In terms of the notation of eqn (4.13) we may then write

$$3\langle\cos^2\zeta\rangle = 2\left\{ S_{y'y'}\sin^2\beta + S_{z'z'}\cos^2\beta + \frac{1}{2}S_{y'z'}\sin 2\beta + \frac{1}{2} \right\} \tag{4.23}$$

Since by definition $\langle\cos^2\zeta\rangle$ is a maximum, its derivative with respect to β must be zero, thus

$$\frac{\partial\langle\cos^2\zeta\rangle}{\partial\beta} = \frac{2}{3}\left\{ \left(S_{y'y'} - S_{z'z'}\right)\sin(2\beta) + S_{y'z'}\cos(2\beta) \right\} = 0 \tag{4.24}$$

which gives eqn (4.15) upon rearranging.

To derive the eqn (4.14) expression for LD^r we write

$$\hat{\boldsymbol{\mu}} = (0, \sin\alpha', \cos\alpha')_{x'y'z'} \tag{4.25}$$

where the final subscript indicates we are using the $\{x', y', z'\}$ axis system. To evaluate the LD^r we require $\langle\hat{\mu}_Z^2\rangle$ and $\langle\hat{\mu}_Y^2\rangle$. Now

$$\hat{\mu}_Z = \mathbf{Z} \cdot \hat{\boldsymbol{\mu}} \tag{4.26}$$

so

$$\langle\hat{\mu}_Z^2\rangle = \left\langle \left\{ (\cos\xi', \cos\psi', \cos\zeta') \cdot (0, \sin\alpha', \cos\alpha') \right\}^2 \right\rangle$$

$$= \langle\cos^2\psi'\rangle\sin^2\alpha' + \langle\cos^2\zeta'\rangle\cos^2\alpha' + 2\langle\cos\psi'\cos\zeta'\rangle\sin\alpha'\cos\alpha'$$

$$= \frac{1}{3}\left\{ 2\left(S_{y'y'}\sin^2\alpha' + S_{z'z'}\cos^2\alpha' + S_{y'z'}\sin\alpha'\cos\alpha'\right) + 1 \right\}$$

$$\tag{4.27}$$

As the orientation is uniaxial, the component of the transition moment that is perpendicular to Z is equally likely to be oriented in any direction so

$$\langle\hat{\mu}_X^2\rangle = \langle\hat{\mu}_Y^2\rangle = \frac{1}{2}\left(1 - \langle\hat{\mu}_Z^2\rangle\right) \tag{4.28}$$

So, from eqn (4.5)

$$LD^r = 3\left(\langle\hat{\mu}_Z^2\rangle - \langle\hat{\mu}_Y^2\rangle\right)$$

$$= \frac{3}{2}\left(3\langle\hat{\mu}_Z^2\rangle - 1\right) \tag{4.29}$$

$$= 3\left(S_{y'y'}\sin^2\alpha' + S_{z'z'}\cos^2\alpha' + S_{y'z'}\sin\alpha'\cos\alpha'\right)$$

the final line of this equation being eqn (4.14). Its simple form explains the rather complicated definitions of the orientation parameters in eqns (4.10) and (4.13).

4.5 Determination of transition polarization

Polarizations of transition moments are needed for determining structural information, for example for probing the binding of the molecule to some

Eqn (4.22) is appropriate for a single molecule—hence no averaging.

$\hat{\boldsymbol{\mu}}$ is a unit vector.

oriented macromolecule, from *LD*. Such information may be obtained from inspection of the molecular structure supported by a simple quantum-mechanical calculation, or may be determined experimentally as outlined below. Most of the molecules that are studied by *LD* have at least some symmetry elements. This restricts the possible polarizations of transition moments to being along rotation axes or perpendicular to reflection planes (or both). We shall therefore begin this section by using symmetry to categorize molecules and their orientation axes before we attempt to determine transition moment polarizations. The point group symmetry label for each category of molecule is given in the margin.

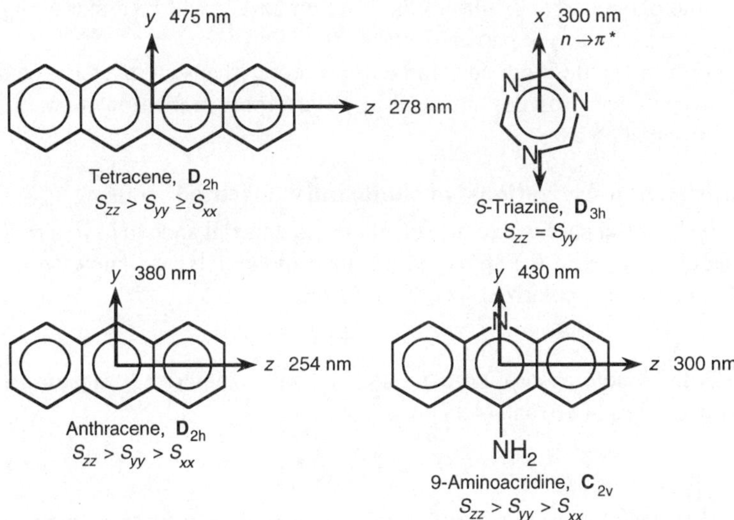

Fig. 4.8 Examples of symmetry-determined molecular orientation axes and transition polarization directions.

(a)　For molecules with tetrahedral or higher symmetry the question of transition moment polarization is irrelevant since all polarizations occur for all transitions. Such molecules will not show *LD* unless they are distorted from their high symmetry geometry.

(a) $\mathbf{T}$, $\mathbf{T_d}$, $\mathbf{T_h}$, $\mathbf{O}$, $\mathbf{O_h}$, $\mathbf{I}$, $\mathbf{I_h}$, and the spherical point groups.

(b)　A molecule with one three-fold or higher rotation axis has transitions polarized either along this rotation axis or perpendicular to it. The polarization of any transition that is in the plane perpendicular to the major rotation axis is delocalized over all directions in the plane. z for such a molecule is either the unique axis in which case $S_{yy} = S_{xx} = -\frac{1}{2} S_{zz}$, or z lies in the plane perpendicular to the unique axis of the molecule and is indistinguishable from y so $S_{zz} = S_{yy} = -\frac{1}{2} S_{xx}$. When z lies in the plane perpendicular to a unique high order symmetry axis then the orientation of the x axis is greatest, despite the fact that it lies perpendicular to the orienting force.

(b) $\mathbf{C_n}$, $\mathbf{C_{nh}}$, $\mathbf{C_{nv}}$, $\mathbf{D_n}$, $\mathbf{D_{nh}}$, $n \geq 3$, $\mathbf{D_{nd}}$, $n \geq 2$, and $\mathbf{S_n}$, $n \geq 4$.

(c)　Molecules with three two-fold rotation axes but no higher order rotation axes and those with one two-fold rotation axis and two reflection planes must have all transitions polarized either along a

(c) $\mathbf{D_2}$, $\mathbf{D_{2h}}$, and $\mathbf{C_{2v}}$.

(d) **C$_s$** and **C$_{2h}$**.

(d) Molecules with only a reflection plane, or a reflection plane and a two-fold axis perpendicular to it, have transitions polarized either in or perpendicular to the plane. Each in-plane transition has a well-defined polarization, but there is no symmetry-based restriction on what polarizations are allowed. As discussed above, z will lie along some well-defined (but not symmetry-determined) axis in the plane.

(e) **C$_2$**.

(e) Molecules with only a two-fold rotation axis have transitions polarized either along the axis or perpendicular to it. As with category (d) molecules there is no restriction on the polarizations of transitions in the plane perpendicular to the rotation axis. z will be either along the symmetry axis or perpendicular to it.

(f) **C$_1$** and **C$_i$**.

(f) For molecules with no symmetry (except perhaps an inversion centre), symmetry provides no help in determining transition moments or orientation axes.

rotation axis or perpendicular to a reflection plane. z is one of the symmetry-determined axes.

Transition polarizations of uniaxially oriented rods

For the simplest case where the orientation is uniaxial and we have a rod-like molecule [*e.g.* case (b) above with z the unique axis] we know from eqn (4.16) that for a z-polarized ($\alpha = 0°$) transition

$$LD^r(z) = +3S_{zz} \tag{4.30}$$

across the whole absorption band and for an x/y polarized transition ($\alpha = 90°$) from eqns (4.16) and (4.17)

$$LD^r(x/y) = -\frac{3}{2}S_{zz} \tag{4.31}$$

It is thus easy to assign transition polarizations simply from the sign of the LD^r as illustrated in Fig. 4.9. LD^r signals of intermediate value (often sloping across an absorption envelope) indicate the presence of overlapping bands of different polarizations.

Despite its low solubility in polyethylene, a drop of a saturated solution of tetracene (Fig. 4.8) in cyclohexane/dichloromethane after a couple of minutes in contact with a stretched polyethylene film produces the LD spectrum shown in Fig. 4.9a. Both positive and negative peaks are apparent in the spectrum showing the presence of in-plane $\pi \to \pi^*$ transitions polarized both parallel (z-polarized) and perpendicular (y-polarized) to the long symmetry axis of the molecule. From this spectrum we can determine both the orientation parameters and the identity of differently polarized overlapping transitions.

As a first approach, let us assume that the strong positive peak at 278 nm is of pure z polarization. Eqn (4.18) and the LD^r can thus be used to yield the orientation parameter for the long axis of the tetracene molecule:

$$S_{zz} = \frac{1}{3}LD^r \approx 0.45 \tag{4.32}$$

The orientation is thus reasonably efficient, as would be expected from the large dimensions of the molecule. By performing the same operation on the LD peak with the most negative LD^r signal, the one at 475 nm, we obtain the value of S_{yy} to be approximately -0.22. From the fact that the orientation parameters sum to zero, we then deduce that $S_{xx} \approx S_{yy}$ as expected for a rod [eqn (4.17)]. We may therefore conclude that our assignments of these band polarizations is correct.

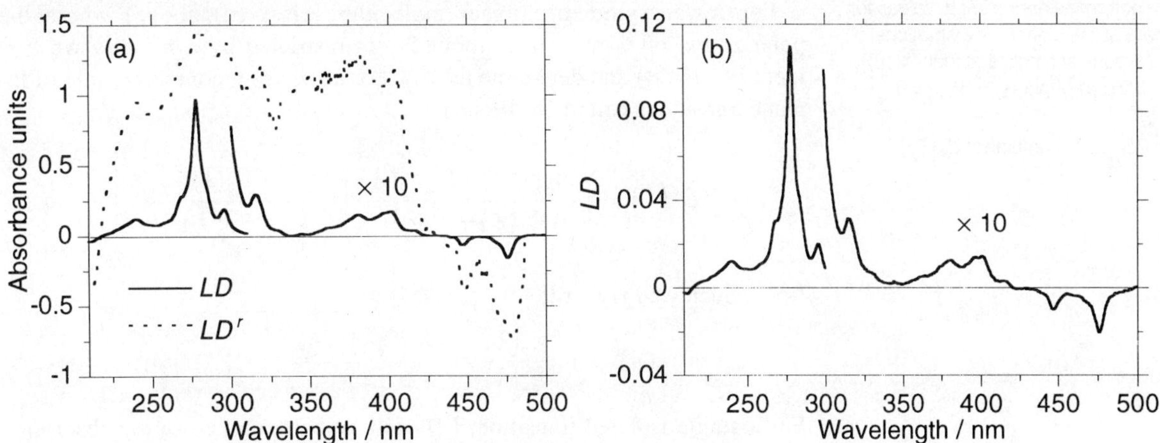

Fig. 4.9 (a) *LD* and *LD'* spectra for the approximately uniaxial rod molecule tetracene oriented in a stretched polyethylene film (2 × stretched sandwich bag, solvent 5% CH_2Cl_2 in hexane) and (b) *LD* spectrum of tetracene in the liquid crystal described in §1.4.[1]

A significant extent of mixing of long axis character into the short wavelength region of the short axis polarized transition at 450 nm is seen by the gradual change of LD^r towards less negative, and below 430 nm, even positive values. The situation for anthracene is much the same, though we ignored this coupling to produce the simplified spectra given in Chapter 1.

If the absorbance is large enough, A may be determined from

$$A = \frac{1}{3}(A_z + 2A_Y)$$

Transition polarizations of planar molecules

It is generally true that planar molecules partially oriented in polymers or liquid crystals have x perpendicular to the plane of the molecule and exhibit a slightly lower orientation for this axis (S_{xx}) than for the in-plane axis perpendicular to z (S_{yy}). The deviation from rod-like behaviour (which may be quantified by the difference $|S_{yy} - S_{xx}|$ called the 'biaxiality') becomes more apparent with shorter molecules. If x, y, and z are defined by symmetry [case (c) above], then the identity of z is usually apparent upon inspection. Alternatively, z can be determined from transitions of known polarization (often vibrational transitions) as was illustrated for riboflavin in §4.4.

The LD^r for pure z-, y-, or x-polarized transitions is given by eqn (4.18):

$$LD^r(z) = 3S_{zz}, \quad LD^r(y) = 3S_{yy}, \quad LD^r(x) = 3S_{xx} \tag{4.33}$$

So, z-polarized transitions can be identified by their large positive LD^r and y-polarized ones by their lower or negative LD^r. In contrast to the situation for rods discussed above, the magnitude of the LD^r for y-polarized transitions is not linked directly to that of z-polarized transitions, though all transitions of the same polarization should have the same values, so mixed polarization bands should be apparent.

For example, the near symmetric drug, quinacrine, when oriented in stretched PVA, has $S_{zz} = 0.33$, but S_{yy} only equals -0.1. The relative importance of the in-plane short axis for the orientation of the molecule is greater in this example than for tetracene. The limit of this behaviour is shown by disk-like molecules such as benzene where $S_{zz} = S_{yy} > 0$.

For lower symmetry planar molecules [class (d) above] where the polarization of in-plane transitions is not restricted by symmetry we first identify z (§4.4) and determine the S_{kk}. For a spectrum composed only of in-plane transitions eqn (4.18) becomes

To determine approximate values for the S_{kk}, consider many vibrational transitions and note that [eqn (4.18)]
$$LD^r(\text{out-of-plane}) = 3S_{xx} < 0$$
and
$$3S_{yy} \leq LD^r(\text{in-plane}) \leq 3S_{zz}$$

$$LD^r(\lambda) = 3\left(\frac{S_{zz}\varepsilon_z(\lambda) + S_{yy}\varepsilon_y(\lambda)}{\varepsilon_z(\lambda) + \varepsilon_y(\lambda)}\right) = 3\left(\frac{S_{zz}\dfrac{\varepsilon_z(\lambda)}{\varepsilon_y(\lambda)} + S_{yy}}{\dfrac{\varepsilon_z(\lambda)}{\varepsilon_y(\lambda)} + 1}\right) \tag{4.34}$$

from which we may write

$$\frac{\varepsilon_z(\lambda)}{\varepsilon_y(\lambda)} = -\frac{LD^r(\lambda) - 3S_{yy}}{LD^r(\lambda) - 3S_{zz}} \tag{4.35}$$

For a single isolated transition, LD^r will be constant across the absorption envelope as will the ratio of z and y extinction coefficients so we may determine the transition polarization for an in-plane transition of a planar molecule from

$$\frac{\varepsilon_z(\lambda)}{\varepsilon_y(\lambda)} = \frac{\cos^2\alpha}{\sin^2\alpha} = \cot^2\alpha \tag{4.36}$$

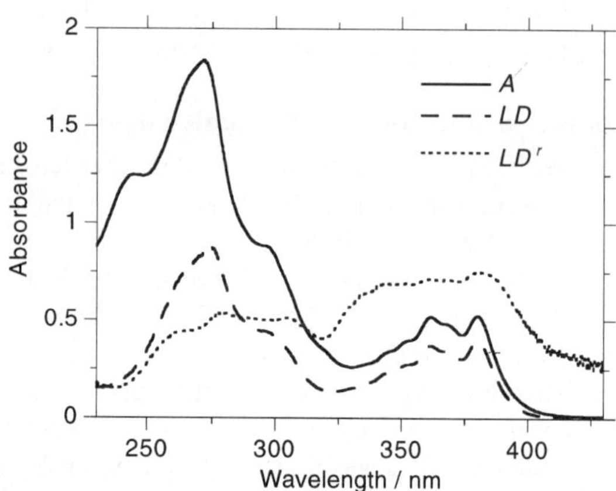

By symmetry any transition in DPPZ should be either y or z polarized, so the LD^r should take either of two values. The fact that in reality the LD^r varies over a wide range of values is the result of overlapping transitions (see Fig. 4.12.)

Fig. 4.10 Isotropic absorbance, *LD*, and *LD*^{r} spectra for the uniaxially oriented planar molecule dipyrido(3,2-a:2',3'-c)phenazine (DPPZ) oriented in a stretched PVA film.[11]

If orientational behaviour is of primary interest, for example in order to characterize the interactions between a solute molecule and the surrounding anisotropic host, it is practical to display the S_{yy} and S_{zz} values in an orientation triangle. The borders and inside of this triangle define all possible combinations of the two largest orientation parameters for any kind of situation and molecule. The lower line, corresponding to $S_{yy} = -0.5S_{zz}$, starting at the coordinate (0,0) and ending at (1.0, –0.5) represents rod-like molecules of increasing ability to align in the medium. The line pointing upwards, corresponding to $S_{yy} = S_{zz} > 0$, starting at (0,0) and ending at (0.25,0.25) describes the behaviour of disks.

Most planar molecules exhibit orientational properties that fall between the two extremes of rod-like and disk-like behaviour. When transition polarizations of a new molecule are required, consideration of where related molecules lie in the orientation triangle may be used to give estimates of the orientation parameters to be used in transition moment polarization assignments.

The degree of alignment is approximately proportional to the length of the molecule.

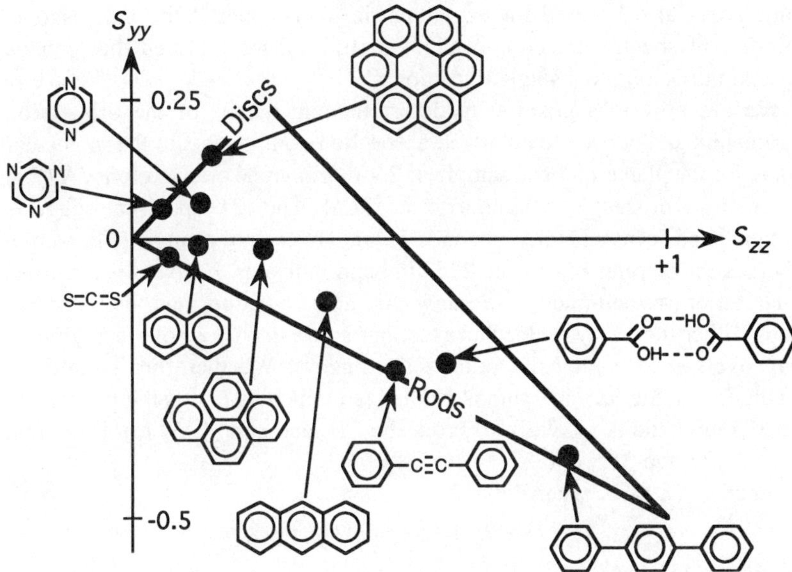

Fig. 4.11 Orientation triangle: plot showing largest (S_{zz}) and intermediate (S_{yy}) orientation parameters for molecules in uniaxial orientation matrices. The molecules, which are all oriented in stretched PE, are depicted so their *z* axes are horizontal.

Determination of polarizations of overlapping transitions

When we cannot isolate a transition from other transitions in that their absorptions overlap we need a way of separating out components. A method based on the principle that for $\pi \rightarrow \pi^*$ transitions of planar molecules [where $A_x(\lambda) \approx 0$] the macroscopic experimental polarized spectra $A_Z(\lambda)$ and $A_Y(\lambda)$ can be deconvoluted into the molecular absorption components $A_z(\lambda)$ and $A_y(\lambda)$ is known as the 'Trial and Error Method' (TEM) invented by Eggers and Thulstrup in the early 1960s without any assumption of a uniaxial sample[12] and developed further together with Michl. For high-symmetry molecules $A_z(\lambda)$ is obtained by identifying the linear combination

$$A_z(\lambda) \propto A_Z(\lambda) - \delta_z A_Y(\lambda) \tag{4.37}$$

which is a spectrum with a 'good' shape, that is everywhere non-negative, and has no feature that for some reason is known to belong to the orthogonal spectrum $A_y(\lambda)$. This is achieved by changing δ in small increments. The $A_y(\lambda)$ spectrum may then be determined in an analogous manner from

$$A_y(\lambda) \propto A_Y(\lambda) - \delta_y A_Z(\lambda) \tag{4.38}$$

TEM can also conveniently be applied by forming linear combinations of $A(\lambda)$ and $LD(\lambda)$.[1]

TEM has successfully been used to provide resolved polarized spectra and transition assignments for numerous polycyclic aromatic molecules oriented in stretched polyethylene films.[5,6] It has thus substantially increased our understanding of the electronic structures and excited states of many organic molecules and has also given important controls for quantum mechanical calculations. TEM is most successful for symmetric molecules but may in some cases also be used for unsymmetric planar ones if the macroscopic component spectra are composed of distinct sharp features that can be assumed to belong to a single transition.

We are now in a position to determine the details of the orientation parameters and hence the polarized absorption components in the molecular frame for the planar uniaxial sample of 2,7-diazapyrene oriented in PVA (Fig. 4.4) using a process very similar to the TEM. The LD^r spectrum suggests that the bands centred at 325 nm and 375 nm are each of a single polarization whereas the sloping LD^r of the 225 nm band indicates this band is of mixed polarization (by symmetry we know that any transition must be polarized along either the z or y molecular axes, but some intensity polarized along z may overlap with intensity polarized along y). We therefore tentatively conclude that the 325 nm band is z-polarized (since its LD^r is the larger) and the 375 nm band is y-polarized. From the LD^r and eqn (4.18) it follows that $3S_{zz} = 0.77$ and $3S_{yy} = 0.22$.

Since $\varepsilon_x(\lambda) = 0$, we may write

$$LD = S_{zz} A_z(\lambda) + S_{yy} A_y(\lambda) \tag{4.39}$$

where $A_z(\lambda)$ is the absorbance of the sample if the radiation is polarized along the molecular axis z *etc.* Thus

$$S_{yy} A_z(\lambda) - S_{zz} A_z(\lambda) = S_{yy} A_y(\lambda) + S_{yy} A_z(\lambda) - LD \tag{4.40}$$

hence

$$A_z(\lambda) = \frac{S_{yy} A_y(\lambda) + S_{yy} A_z(\lambda) - LD}{S_{yy} - S_{zz}}$$
$$= \frac{3 S_{yy} A(\lambda) - LD}{S_{yy} - S_{zz}} \tag{4.41}$$

where $A(\lambda)$ is the isotropic absorbance at λ and similarly

$$A_y(\lambda) = \frac{S_{zz} A_z(\lambda) + S_{zz} A_y(\lambda) - LD}{S_{zz} - S_{yy}}$$
$$= \frac{3 S_{zz} A(\lambda) - LD}{S_{zz} - S_{yy}} \tag{4.42}$$

If our assumptions about the pure z and y polarizations of the transitions at 325 nm and 375 nm are indeed correct then the $A_z(\lambda)$ and $A_y(\lambda)$ spectra calculated according to eqns (4.41) and (4.42) using our calculated orientation parameters should be the correct component spectra. The final calculated polarized spectra should be scrutinized with respect to consistency: they should not be negative for any λ (within experimental error) and the two polarized spectra should 'look different', *i.e.* not have any vibrational or other specific features in common. The component spectra are illustrated in Fig. 4.12a.

Analogously resolved component spectra of tetracene (Fig. 4.9) and DPPZ (Fig. 4.10) are also shown in Fig. 4.12. The DPPZ resolved absorption envelopes confirm the existence of a y-polarized transition with origin at 375 nm and vibronic components continuing below 400 nm, a second z-polarized transition at around 400 nm, with absorption overlapping the first one, and a strong z-polarized transition below 300 nm. These results are in fair agreement with predicted energy positions and polarizations from simple PPP calculations. A weaker, y-polarized transition (predicted at about 300 nm) may explain the broad negative component indicated at 330 nm.

The idealized *LD* spectrum with $S = 0$ for diaminopyrene illustrated in Fig. 4.4 was constructed from these component spectra.

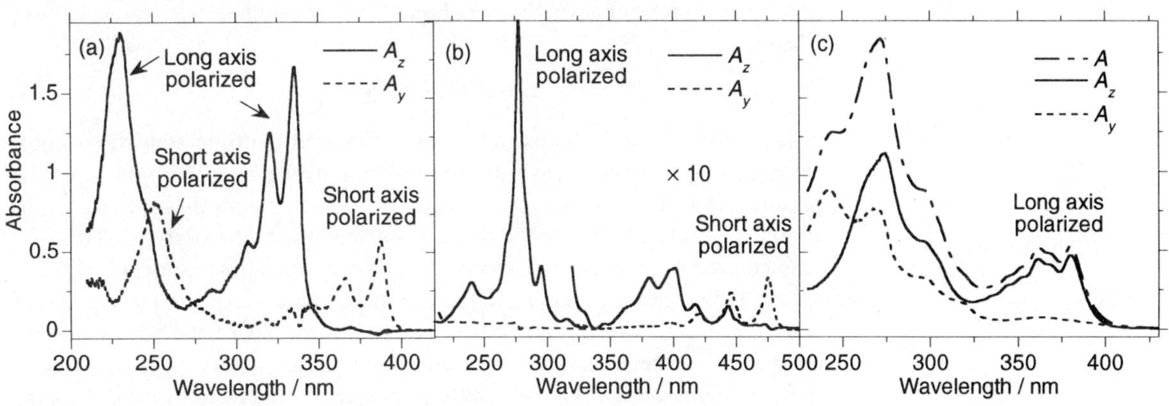

Fig. 4.12 Polarized absorption spectra of (a) diazapyrene (Fig. 4.4), (b) tetracene (Fig. 4.9), and (c) DPPZ (Fig. 4.10) determined as described in the text.

With broad, strongly overlapping bands such as in the nucleobases, purines, indoles, flavins, and many other biologically important chromophores, it is not so easy to determine 'good' component spectra. An alternative method for this situation[10,14] has been used for a wide variety of planar chromophores and some macromolecules.[13,17] It consists of the following steps:

(1) *Determine S_{zz}, S_{yy}, and the z-axis by an independent method.* For example, measure the infra-red *LD* of three transitions with known directions and analyse the spectra as discussed in §4.4.

(2) *Decompose $A(\lambda)$ into (e.g. Gaussian) components $A_i(\lambda)$ each corresponding to a transition.*

(3) *Fit the LD^r (or LD) spectrum* using these components and the two S parameters to determine the α value for each transition as described above.

(4) *Iteratively adjust* shapes and positions of the $A_i(\lambda)$ and the angles α_i to optimize the fit of $A(\lambda)$ and $LD(\lambda)$.

(5) *Resolve sign ambiguities* (is it $+\alpha_i$ or $-\alpha_i$?) either by comparing chromophores that are spectroscopically similar but that have different orientations or by measuring the fluorescence anisotropy (see below).

Fluorescence anisotropy

Fluorescence anisotropy (FA) is defined for fluorescence intensities measured with polarizers in both the excitation and emission beams of the fluorimeter. The excitation polarizer is oriented so as to vertically polarize the light and the emission polarizer oriented first to transmit vertically polarized light and then horizontally polarized light in the emission beam (Fig. 4.13):

In practice, to correct for inherent polarization effects of the instrument one should use[16]

$$FA = \frac{I_{VV} - GI_{VH}}{I_{VV} + 2GI_{VH}}$$

where

$$G = \frac{I_{HV}}{I_{HH}}$$

$$FA = \frac{I_{VV} - I_{VH}}{I_{VV} + 2I_{VH}} \tag{4.43}$$

where, *e.g.*, I_{VH} is the emerging intensity of light appearing with a vertical excitation polarizer and a horizontal emission polarizer.

Let us assume for simplicity that there is no loss of polarization due to rotation of the chromophore during the lifetime of the emission process or to energy transfer between chromophores. For an isolated transition we may then write:[18]

$$FA = \tfrac{1}{5}\left(3\cos^2 \chi - 1\right) \tag{4.44}$$

where χ is the angle between the absorbing and emitting transition dipole moments. Whichever transition we choose to excite into, the light is always emitted from the lowest excited state (of the same spin multiplicity as the initial excited state, assuming there is no singlet-triplet crossings). Thus, FA can be used to give the angles *between* different transition moments.

FA is the fluorescence analogue of the LD^r and is expected to equal $+\tfrac{2}{5}$ for transitions whose polarization is parallel to that of the lowest energy transition, since then $\chi = 0$. Its minimum value is $-\tfrac{1}{5}$, corresponding to perpendicular absorbing and emitting moments. To account for overlapping transitions a sum over transitions must be included in eqn (4.44).

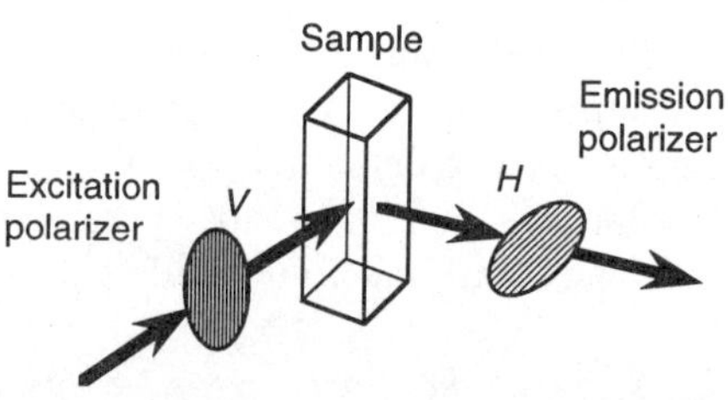

Fig. 4.13 *FA* experiment with vertical excitation and horizontal emission polarizer settings (light source, detector, and monochromators omitted for clarity).

Resolution of angle-sign ambiguities

As noted above the use of LD to determine transition polarizations almost always results in two possible values for α. We shall therefore conclude this section with a brief look at how this ambiguity may be resolved.[10] One option is to use *FA*. The potential of *FA* to aid in interpretation of an *LD* spectrum is due to the fact that it can be used to give the angle *between* transition moments. For example, consider a molecule with two transitions that have the same $\cos^2\alpha$ value. There are four possible sets of polarizations for their transition moments (unless $\alpha = 0$ or $90°$).

- Both transitions are on the same side of z with $\alpha > 0$.
- Both transitions are on the same side of z with $\alpha < 0$.
- The transitions are on opposite sides of z, the first with $\alpha > 0$ and the second with $\alpha < 0$.
- The transitions are on opposite sides of z, the first with $\alpha < 0$ and the second with $\alpha > 0$.

FA, by probing $|\alpha_i - \alpha_1|$, will distinguish some of these possibilities.

Alternatively, the so-called 'substituent-perturbed orientation method' may be used to distinguish between the two mathematical solutions to the $\cos^2\alpha$ problem since we usually need to know on which side of the molecular orientation axis z the transition moment lies. Upon introducing into the chromophore molecule a bulky substituent at a suitable position that does not coincide with the line of the orientation axis, the orientation axis will be rotated towards the substituent since it is in this direction that the solute molecule has been made longer. Thereby, angles between the orientation axis and such transition moments that lie on one side of the old orientation axis will be larger, whereas angles on the other side will be smaller, thus allowing the correct α value for the original molecule to be identified.

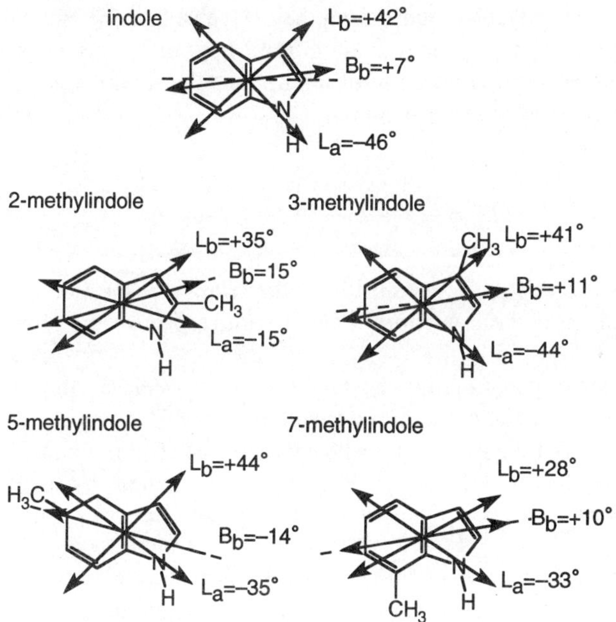

Fig. 4.14 Illustration of the substituent-perturbed orientation method. Dashed lines indicate molecular orientation axes.[14,20]

The substituent-perturbed orientation method has worked well for a number of planar but otherwise unsymmetric hetero-atomic chromophores, and has shown good agreement with quantum mechanical calculations which are, thus, an alternative way of deciding between the two signs of cos α. That the substituent generally, in addition to changing the orientation of the solute molecule in the anisotropic host, also induces an electronic perturbation which may change the transition moment polarizations is a problem that may be avoided by choosing substituents with large orientational but small electronic perturbations.[10,14,17,20]

4.6 *LD* of flow-oriented polymers

In Chapter 3 we considered the effect of orienting the long polymer DNA and how to use the resulting data. To conclude this chapter we shall briefly consider this type of experiment a little more rigorously. The simplest orientation situation for a polymer is that of uniaxial orientation (for example, in an electric field) in which case the (single) orientation parameter depends only on the angle ζ between the polymer-fixed orientation axis and the field. In the commonly used streaming devices of Couette flow type (§1.4), however, the orientation is generally not uniaxial, but biaxial, so S is instead a function of two angles.

The LD^r can be still factorized into an optical factor, O, and an orientation factor, S, as in eqn (4.7) provided there is a random distribution of the transition moment around some polymer-fixed axis (in DNA the local helix axis). Because of the rather complicated functional dependence of S on polymer stiffness, solvent viscosity, flow speed *etc.*, it is often more readily, and more accurately, determined from the LD^r value of some internal standard than from a non-empirical calculation. Including the possibility of overlapping absorption bands of different transitions, which is a rule rather than an exception with most biopolymers, we have the following general formula:

$$LD^r = \frac{3}{2}S\left\{\frac{\sum_i A_i(\lambda)\left(3\cos^2\alpha_i - 1\right)}{\sum_i A_i(\lambda)}\right\} \tag{4.45}$$

with A_i and α_i denoting for transition i the relative absorption contribution and the angle of the transition moment with respect to the local polymer axis.

In the case of pure (non-overlapping) transitions we have eqn (3.1) for each transition. For example, the LD spectrum of the DNA complex with methyl green shows that the transition responsible for the absorption at 630 nm has $\alpha \approx 45°$ and that at 425 nm $\alpha \approx 90°$, if S is 'calibrated' by assuming that the DNA bases give rise to an LD described by an effective polarization $\alpha_{\mathrm{eff}} \approx 80°$.[19]

In some polymers it is of interest to analyse the orientation of subunits which contain chromophores. For example, returning to the example of DNA orientation considered in Chapter 3, we might wish to determine the orientation of DNA bases from the LD^r of flow-oriented DNA. In this case, Z is the helix axis. Let us define the DNA base orientation (Figs 3.1 and 3.3) in terms of tilt, θ_X, around the dyad axis of a base pair and roll, θ_Y, around the axis perpendicular to the dyad axis and let each base have a set of transition moments polarized at angles δ_i relative to dyad axis in the base-plane:

$$LD^r = 3S \frac{\sum_i \frac{3}{2} A_i(\lambda)\left(-\sin\delta_i \sin\theta_X + \cos\delta_i \cos\theta_X \sin\theta_Y\right)^2}{\sum_i A_i(\lambda)} \qquad (4.46)$$

The corresponding relation for out-of-plane polarized transitions for which $\alpha = 0°$ is:

$$LD^r = \frac{3}{2} S\left\{\frac{\sum_i A_i(\lambda)\left(3\cos^2\theta\cos^2\rho - 1\right)}{\sum_i A_i(\lambda)}\right\} \qquad (4.47)$$

Eqn (4.45) and its analogues have also been used to determine orientation of DNA bases within a polymer.[13,15]

References

(1)   Becker, H.-C. *Unpublished data*

(2)   Nordén, B. *App. Spectrosc. Rev.* **1978**, *14*, 157; Nordén, B.; Kubista, M.; Kuruscev, T. *Q. Rev. Biophys.* **1992**, *25*, 51

(3)   Schellman, J. A.; Jensen, H. P. *Chem. Rev.* **1987**, *87*, 1359.

(4)   Wada, A. *Appl. Spectrosc. Rev.* **1972**, *6*, 1

(5)   Samori, B.; Thulstrup, E. W., (Eds.); *Polarized spectroscopy of ordered systems.* The Netherlands: Kluwer Academic Publishers, **1988**

(6)   Michl, J.; Thulstrup, E. W. *Spectroscopy with polarized light.* New York: VCH, **1986**.

(7)   Fredericq, E.; Houssier, C. *Electric dichroism and electric birefringence.* Oxford: Clarendon Press, **1973**

(8)   Charney, E. *Q. Rev. Biophys.* **1988**, *21*, 1

(9)   Breton, J.; Verméglio, A., (Eds.) *The photosynthetic bacterial reaction centre: structure and dynamics.* New York and London: Plenum Press, **1988**

(10) Matsuoka, Y.; Nordén, B. *J. Phys. Chem.* **1982**, *86*, 1378; **1983**, *87*, 220

(11) Lincoln, P.; Broo, A.; Nordén, B. *J. Amer. Chem. Soc.* **1996**, *118*, 2644

(12) Eggers, J.H.; Thulstrup, E. W. *Eighth european congress on molecular spectroscopy.* Copenhagen, **1965**; Thulstrup, E. W.; Michl, J.; Eggers, J. H. *J. Phys. Chem.* **1970**, *74*, 3868; 3878

(13) Matsuoka, Y.; Nordén, B. *Biopolymers* **1982**, *21*, 2433.

(14) Albinsson, B.; Nordén, B. *J. Phys. Chem.* **1992**, *96*, 6204

(15) Chou, P-J.; Johnson, C. J. Jr. *J. Amer. Chem. Soc.* **1993**, *115*, 1205

(16) Lacowitz, J. R. *Principles of fluorescence spectroscopy.* New York: Plenum Press, **1983**

(17) Albinsson, B.; Kubista, M.; Sandros, K.; Nordén, B. *J. Phys. Chem.* **1990**, *94*, 4006

(18) Cantor, C. R.; Schimmel, P.R. *Biophysical chemistry. Part II. Techniques for the study of biological structure and function.* San Fransisco: Freeman and Co., **1980**

(19) Nordén, B.; Tjerneld, F. *Chem. Phys. Lett.* **1977**, *50*, 508

(20) Albinsson, B.; Kubista, M.; Thulstrup, E.; Nordén, B. *J. Phys. Chem.* **1989**, *93*, 6646

5 Analysis of circular dichroism: electric dipole allowed transitions

5.1 Introduction

CD probes optical activity of electronic transitions if we use UV/visible light. If infra-red radiation is used then we measure vibrational optical activity.

The absorption spectrum of a solution of chiral molecules measured with left circularly polarized radiation would differ by at the very most 2% from that measured with unpolarized or right circularly polarized light. For most samples the difference between the absorbances measured with left and right circularly polarized light rarely exceeds ±0.001%. A *CD* spectrum is thus the result of a small difference between two large absorbance numbers. It gives only the helical or asymmetric part of the change that occurs when radiation is absorbed and this is probed by a combination of the electric and magnetic fields of the radiation. The smallness of the difference is due to the fact that at a given field strength magnetic field interaction energies are some thousand times weaker than the corresponding electric field interaction energies.

At least in principle, a *CD* spectrum contains all the information we might wish to know about the asymmetry of the system we are studying. In reality, how *CD* is used and interpreted varies considerably depending on the systems studied. Whilst by no means always the case, it is often true that large molecular systems are analysed qualitatively or empirically as we saw for biomolecules in Chapter 2, whereas small systems are analysed in more quantitative ways with a more secure theoretical foundation.

In order to interpret a *CD* spectrum in geometric terms it is important that we understand how the signal arises. One reason *CD* has only very gradually been adopted as a structural tool is that there is a 50% chance of correctly guessing which enantiomer is in the solution, and it has not always been obvious that the success rate has been improved by using *CD*. The aim of this chapter is to provide a sound basis for relating *CD* and geometry. After a general discussion of what *CD* is (§5.2) and the ways a spectrum can be analysed (§5.3) we shall focus on the *CD* of electric dipole allowed transitions. Magnetic dipole allowed transitions form the subject matter of the next chapter. The equations used in both chapters are derived in Chapter 7.

5.2 Pictorial description of *CD* spectroscopy

An electronic transition occurs because either the electric field or the magnetic field (or both) of the radiation 'pushes' the electrons to a new stationary state. The effect of the electric field is to cause a linear rearrangement of the electrons; the net linear displacement of charge during any transition is therefore called the *electric dipole transition moment* (edtm) of the transition and is denoted by the vector $\boldsymbol{\mu}$. The direction in which $\boldsymbol{\mu}$ points is the same as the transition polarization—it is the direction in which the electrons are pushed during the transition. In contrast to the linear effect of the electric field, the magnetic field induces a circular rearrangement of electron density as illustrated in Fig. 5.1. The net circulation of charge is the *magnetic dipole transition moment* (mdtm), $\boldsymbol{m}$.

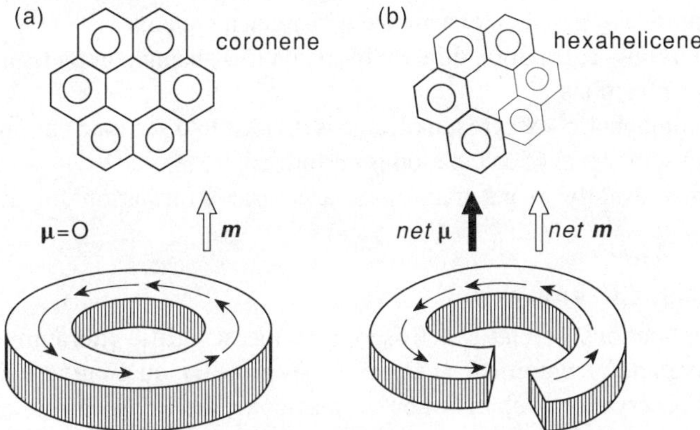

We use bold face type, *e.g.* $\boldsymbol{v}$, to denote vectors; the same symbol in normal face type, v, is the length or magnitude of the vector; a vector of length 1 unit along the direction of a vector is denoted by $\hat{\boldsymbol{v}}$. The components of the vector are denoted by subscripts, so an edtm is

$$\boldsymbol{\mu} = (\mu_x, \mu_y, \mu_z).$$

Fig. 5.1 Electron redistribution upon light excitation for a transition of (a) an achiral molecule and (b) a chiral molecule. The edtms are represented by black arrows and the mdtms by white arrows. The choice of direction of the 'current' dipoles, *i.e.* → or ← , is arbitrary and can be seen as indicating the polarization at a certain time. The same result ($\boldsymbol{\mu}$ parallel to $\boldsymbol{m}$) is obtained for the right-handed helicene helix irrespective of choice of current direction.

$\boldsymbol{\mu}$ is known as a *polar* vector: it has magnitude and direction. $\boldsymbol{m}$ describes rotation about an axis and is known as an *axial* vector. Its direction is such that when viewed from its tip the rotation is anti-clockwise. Alternatively, if one aligns the right-hand thumb along $\boldsymbol{m}$ and holds the fingers in a curve, the circulation of charge follows the fingers.

In an achiral molecule the net electron redistribution is always planar. It is usually linear (having $\boldsymbol{\mu} \neq 0$, $\boldsymbol{m} = 0$) or circular (having $\boldsymbol{\mu} = 0$, $\boldsymbol{m} \neq 0$), but it may have both $\boldsymbol{\mu} \neq 0$ and $\boldsymbol{m} \neq 0$ in which case the electrons rearrange in a planar spiral. In a chiral molecule the electron rearrangement during a transition is always helical. The helix of electron motion results from $\boldsymbol{\mu}$ and $\boldsymbol{m}$ having parallel polarizations and combining to force a helix of electron redistribution. If the helix of electron motion is right-handed then it is more easily induced by left circularly polarized light (of the appropriate frequency) than by right circularly polarized light and we observe a positive *CD* signal. The fact that left-handed light induces a right-handed transition better than a left-handed one follows from the way the electric field vector of the light interacts with $\boldsymbol{\mu}$ and the complex component of the magnetic field simultaneously interacts with $\boldsymbol{m}$. It is not obvious upon inspection.

This result follows from the analysis of Chapter 7.

5.3 Ways of analysing *CD*

Our conclusion from the previous section is that we would expect to see a *CD* signal for a collection of chiral molecules if the photons incident upon the sample have the correct energy to cause a transition. In a *CD* experiment the molecules are typically randomly oriented otherwise *LD* effects (see previous chapters) usually dominate the *CD* signal. The equation that summarizes the requirement of parallel μ and m for the helical electron displacement required for *CD* is the Rosenfeld equation for the *CD* intensity (or *CD* strength or rotational strength, *R*) of a transition in a collection of randomly oriented chiral molecules:

$$R = \mathrm{Im}\{\boldsymbol{\mu} \cdot \boldsymbol{m}\} \tag{5.1}$$

where 'Im' denotes 'imaginary part of', μ is the edtm for the transition from the final to the initial state, and m is the mdtm for the reverse transition.

Methods of analysis of *CD* spectra may be categorized as follows:

- empirical—based on experience with related systems
- *ab initio*—calculation directly from eqn (5.1) using complete molecular wavefunctions
- chromophoric—where a molecule is divided into separate chromophores and some level of calculation is performed.

The aim is usually to get maximum structural information for minimum effort.

Empirical *CD* analysis

Most applications of *CD* spectroscopy in the scientific literature involve qualitative and/or empirical analyses of spectra. By qualitative we simply mean an observation such as 'when … was done, the *CD* spectrum changed, therefore we concluded that … had happened'. For example, the loss of structure in an RNA molecule upon heating may be followed using *CD* (Fig. 5.2). Alternatively, the proof of a chiral (or at least enantioselective) synthesis or separation may be acquired using *CD* (Fig. 5.3). Convergence of *CD* amplitude upon repeated purification is then a criterion of pure enantiomers.

The magnetic dipole moment operator contains the factor $\sqrt{(-1)}$ and hence can give rise to complex functions (see Chapter 7).

The strict definition of a *chromophore* is a sub-unit of a system whose wavefunctions have *no* overlap with the rest of the system; electronic wavefunctions in a chromophore therefore have no electron exchange with the rest of the system. In practice for *CD*, a chromophore is usually identified as a moiety within a molecule whose normal absorption is more or less independent of the rest of the molecule. Chromophores of a molecule can usually be identified with functional groups.

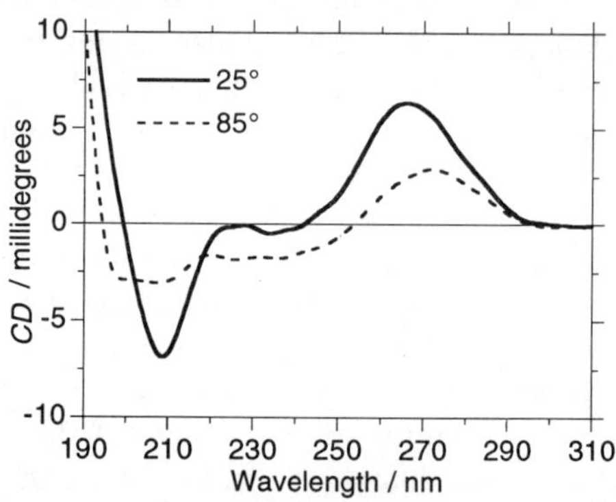

Fig. 5.2 *CD* of a long (372 nucleotide) naturally occurring mRNA. At 25°C the RNA is largely that of an A-form double helix; at 85°C there is less chirality in the parts of the molecule being probed—in this case the RNA bases. This is due to there being less base pairing and base stacking at 85°C.[1] *A*(260 nm) = 1.0 at 25°C.

If the analysis is quantified by measuring the change in the *CD* as a function of concentration, temperature, ionic strength, *etc.*, and/or by comparison with 'related' systems (as is the case for protein structure determination, §2.2) then we describe the analysis as empirical. When using 'related' systems to interpret a new *CD* spectrum it is vital that the systems are related spectroscopically as well as by molecular and electronic structure.

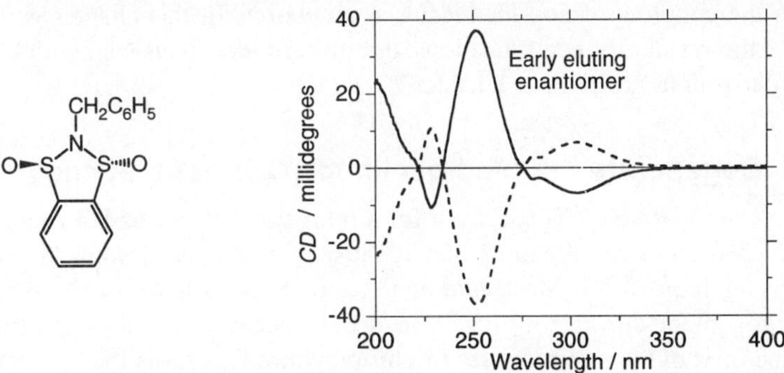

Fig. 5.3 *CD* spectra in acetonitrile for two 1,3,2-benzodithiazole-*S*-oxide enantiomers eluting successively off a chiral HPLC column, thus confirming the successful resolution of the enantiomers from a racemic mixture by HPLC.[2] The 1*R*,3*R* enantiomer, which is illustrated, is eluted from the column first.

Ab initio CD analysis

Ab initio calculations have occasionally been performed on particular systems of interest. However, a great deal of work is required to ensure the calculation is reliable, and the results usually only relate to the single system for which the calculation has been performed. In some instances *ab initio* calculations on a particular class of compounds have proved useful in a wider context. This is certainly true of the calculations of Hansen *et al.* for β-adamantanones (Fig. 1.2) where the calculation results were related back to underlying *CD* mechanisms and so a particular calculation was used to advance our understanding of the theory of *CD*.[3]

Chromophoric analysis of *CD* spectra

The key feature of a chromophore, as of a functional group, is that it can be considered as a spectroscopically well-defined sub-unit of a molecule that is only slightly perturbed by the rest of the system. We shall limit our consideration to achiral chromophores. Thus the isolated chromophores have no intrinsic *CD* and we will concentrate on deducing the *CD* induced into a particular transition by the chromophore's environment. Given that the chromophores are achiral, then either $m = 0$, or $\mu = 0$, or they are perpendicular to one another. The $m = 0$, $\mu \neq 0$ transitions are called *electric dipole allowed* (eda), *magnetic dipole forbidden* (mdf); these form the subject matter of the remainder of this chapter. The $m \neq 0$, $\mu = 0$ transitions are called *electric dipole forbidden* (edf), *magnetic dipole allowed* (mda); they are analysed in the next chapter.

The significance of this division is due to the way *CD* intensity is induced into the two different types of transitions. Eda transitions require the induction of a magnetic component, whereas mda transitions require an induced electric component. The dependence of the different kinds of induced moments on the geometry of the system is very different, so when we wish to extract geometric information from *CD* we must be aware of which situation we have. The *CD* spectrum induced into eda transitions is also different if it arises from the coupling of identical chromophores rather than from the coupling of non-identical chromophores. In this chapter we shall quote the results for eda transitions and discuss their physical significance. The derivations are given in Chapter 7.

5.4 Degenerate coupled-oscillator *CD*: general case

We use the labels **A** and **C** to denote the chromophore of the coupling system. **A** is an achiral chromophore where the transition of interest is located and **C** is the chromophore that provides the chiral perturbation to **A** transitions. In the degenerate coupled-oscillator case **A** also provides a chiral pertubation to **C**.

If we can measure a *CD* spectrum for a transition that is eda but (within its own chromophore, **A**) mdf then it must have acquired some magnetic character [eqn (5.1)]. Most commonly, the required helix of the electron rearrangement during the transition occurs because an eda transition in another part of the molecule, μ^c in chromophore **C**, causes the electrons in the transition of interest, μ^a, to be helically 'deflected'. The induced magnetic character arises since the linear motion of electrons in **C** makes a tangent to a circle about **A** (Fig. 5.4b). Conversely, μ^a gives magnetic character to μ^c.

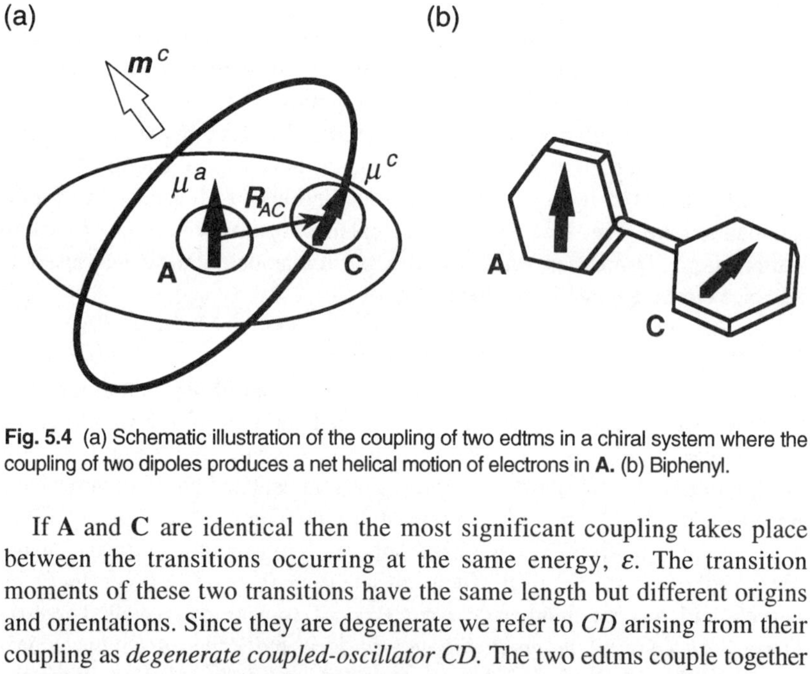

(a)   (b)

If μ^a and μ^c are perpendicular, then they do not interact and no *CD* signal is observed.

Fig. 5.4 (a) Schematic illustration of the coupling of two edtms in a chiral system where the coupling of two dipoles produces a net helical motion of electrons in **A**. (b) Biphenyl.

If **A** and **C** are identical then the most significant coupling takes place between the transitions occurring at the same energy, ε. The transition moments of these two transitions have the same length but different origins and orientations. Since they are degenerate we refer to *CD* arising from their coupling as *degenerate coupled-oscillator CD*. The two edtms couple together to make one of two helices depending upon whether they are in phase or out of phase—so we expect to see a *CD* spectrum with two peaks. As the two edtms that couple are degenerate, the two resulting helices of electron motion have equal magnitude and opposite sign *CD* signals at energies very close to ε so we see the overlapping spectrum of Fig. 5.5. In this section we shall write down the equations that describe the magnitude and sign of the *CD* from the coupling of eda transitions in two distinct identical achiral chromophores, **A** and **C**, then see what sort of spectra are predicted for a number of different cases and compare the results with experimental data.

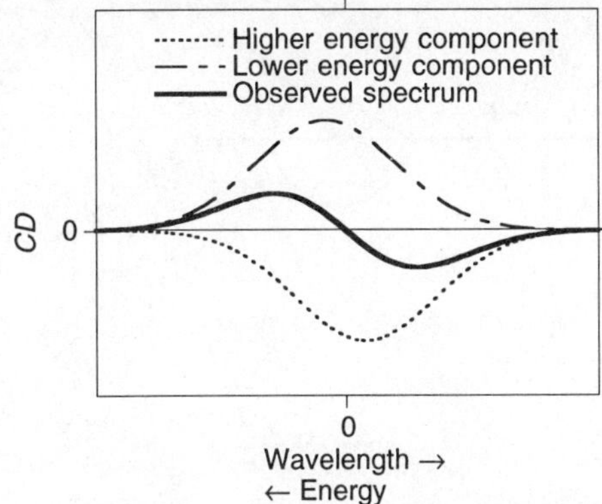

Fig. 5.5 *CD* resulting from the coupling of one transition in each of two chromophores where both transitions are eda and degenerate (occurring at energy ε). The characteristic form of an excitonic spectrum results from cancellation of overlapping positive and negative bands.

If we took the mirror image of the **A/C** system illustrated in Fig. 5.4, then the *CD* spectrum would be the inverse of the one illustrated in Fig. 5.5. It is therefore crucially important that we have a protocol for defining the geometry of the **A/C** system when we wish to determine *CD* arising from the coupling of the edtms on **A** and **C**, μ^{a} and μ^{c}, respectively.

Wherever possible throughout this book we shall illustrate the **A** and **C** chromophores as in Fig. 5.6 with **A** in front and **C** behind. The vector from the **A** origin to the **C** origin, R_{AC}, then goes back into the page from **A** to **C**. In most of the examples we shall look at, we shall define the x-axis to lie along R_{AC}, and z to lie along the y-z projection of μ^{a}. It is also convenient to define three angles: α ($0 \leq \alpha \leq 180°$) the angle between R_{AC} and μ^{a}, γ ($0 \leq \gamma \leq 180°$) the angle between $-R_{AC}$ and μ^{c}, and τ the angle passed through in going from the y-z *projection* of μ^{a} to that of μ^{c} in an *anticlockwise* direction as illustrated in Fig. 5.6. Thus

$$R_{AC} = R_{AC}(1,0,0) \tag{5.2}$$

$$\mu^{a} = \mu^{a}(\cos\alpha, 0, \sin\alpha) \tag{5.3}$$

$$\mu^{c} = \mu^{c}(-\cos\gamma, \sin\gamma\sin\tau, \sin\gamma\cos\tau) \tag{5.4}$$

and

$$\cos\tau = \frac{\hat{\mu}^{a}\cdot\hat{\mu}^{c} - \hat{\mu}^{a}\cdot\hat{R}_{AC}\hat{R}_{AC}\cdot\hat{\mu}^{c}}{\sqrt{\left[1-\left(\hat{\mu}^{a}\cdot\hat{R}_{AC}\right)^{2}\right]\left[1-\left(\hat{\mu}^{c}\cdot\hat{R}_{AC}\right)^{2}\right]}} \tag{5.5}$$

$$= \frac{\hat{\mu}^{a}\cdot\hat{\mu}^{c} + \cos\alpha\cos\gamma}{\sin\alpha\sin\gamma}$$

Any representation of the system is equally valid, but if $0° < \tau < 180°$, the three vectors $\boldsymbol{\mu}^c$, $\boldsymbol{\mu}^a$, and $\boldsymbol{R}_{AC}$ (CAR) form a right-handed coordinate system. If your thumb and two forefingers can lie along the vectors (*cf.* Fig. 1.3) in the CAR order then $0° \le \tau < 180°$. If, however, you would have to break some bones to do it with your right hand, but could use your left hand, then $180° < \tau < 360°$.

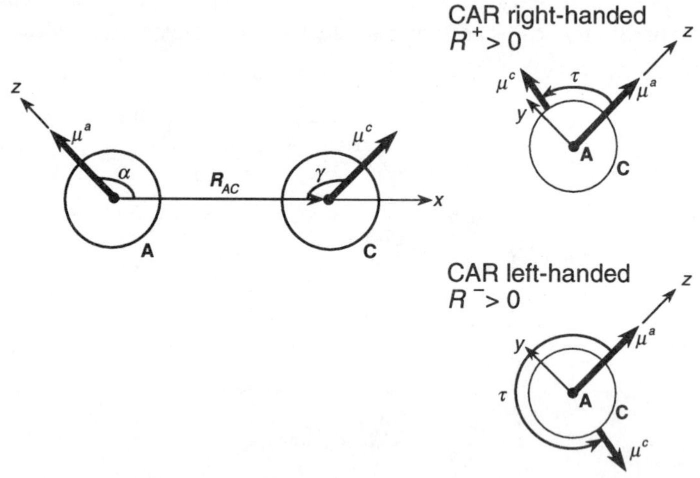

Fig. 5.6 Diagram illustrating the geometry and coordinates for the **A/C** system described in the text. Note that τ is the angle taken in the anticlockwise direction between the *projections* of the edtms onto the *y-z* plane when the observer is looking down the *x*-axis. $0° < \tau < 180°$ if $\boldsymbol{\mu}^c \times \boldsymbol{\mu}^a \cdot \boldsymbol{R}_{AC} > 0$ (*i.e.* if the three vectors form a right-handed parallelepiped) and $180° < \tau < 360°$ if $\boldsymbol{\mu}^c \times \boldsymbol{\mu}^a \cdot \boldsymbol{R}_{AC} < 0$.

Degenerate coupled-oscillator *CD* equations

In order to analyse *CD* spectra in geometric terms we need to have equations relating the spectra to the geometry of the system. These are derived in Chapter 7 and the results summarized here. Two alternative representations of the equations are given: one using vectors and vector products, the other using components of vectors and the angles defined in Fig. 5.6. As illustrated in Fig. 5.5, when degenerate transitions on two chromophores couple we find two *CD* bands of opposite sign that occur at energies slightly above and slightly below the transition energies of the isolated chromophores at energies

The notation $\varepsilon^{\pm}$ means that ε^+ follows the upper of the two signs when there is a choice written $\pm$ or $\mp$ in an equation. So *e.g.* from eqn (5.6) $\varepsilon^+ = \varepsilon + V$.

$$\varepsilon^{\pm} = \varepsilon \pm V \tag{5.6}$$

where

$$V = \frac{\boldsymbol{\mu}^a \cdot \boldsymbol{\mu}^c - 3\hat{\boldsymbol{R}}_{AC} \cdot \boldsymbol{\mu}^a \boldsymbol{\mu}^c \cdot \hat{\boldsymbol{R}}_{AC}}{R_{AC}^3} \tag{5.7}$$

The two *CD* bands have sign and magnitude given by the following vectorial equation (*cf.* Appendix 1)

$$R^{\pm} = \pm \frac{\varepsilon}{4\hbar} \left\{ \boldsymbol{\mu}^c \times \boldsymbol{\mu}^a \cdot \boldsymbol{R}_{AC} \right\} \tag{5.8}$$

where R^+ refers to the transition that occurs at energy ε^+ and R^- to the one that occurs at energy ε^- (*cf.* Fig. 5.5).

Eqn (5.8) describes the magnitude of the area under each *CD* band resulting from the coupling of $\boldsymbol{\mu}^a$ and $\boldsymbol{\mu}^c$; the actual appearance of the spectrum is also dependent on the band shape, which is usually Gaussian or Lorentzian so we get spectra as illustrated throughout this book rather than single sharp lines at precise energies.

From eqns (5.6) and (5.8) we can now qualitatively describe the CD spectrum we expect to see for the coupling of two degenerate transitions on two chromophores.

- There are two bands of equal magnitude and opposite sign, R^+ centred at ε^+ and R^- centred at ε^-.
- The CD strength R depends only on the strength of the transitions in the isolated chromophores and on their orientations relative to each other and to the vector connecting the two chromophores.
- The sign of the CD of the + state is positive if $\boldsymbol{\mu}^c$, $\boldsymbol{\mu}^a$, and $\boldsymbol{R}_{AC}$ form a right-handed axis system.
- V is small, so the two bands are centred close to one another and a significant part of the total intensity of each band is cancelled. Hence the spectral form illustrated in Fig. 5.5.

Mnemonic: CAR is right-handed.

In some applications it is easier to have eqns (5.6) and (5.8) written in terms of angles using the right-handed coordinate system $\{x, y, z\}$ illustrated in Fig. 5.6 and given by eqns (5.2)–(5.5). The degenerate coupled-oscillator CD strength may then be written:

$$R^{\pm} = \pm \frac{\varepsilon \mu^2 R_{AC}}{4\hbar} \sin \alpha \sin \gamma \sin \tau \tag{5.9}$$

and the energies of the two bands are

$$\varepsilon^{\pm} = \varepsilon \pm \frac{\mu^2 \left(\sin \alpha \sin \gamma \cos \tau + 2 \cos \alpha \cos \gamma\right)}{R_{AC}^3} \tag{5.10}$$

Here and subsequently energies and CD magnitudes will be given in unspecified units. The units required for quantitative calculations are given in Appendix 4.

A little care is needed in applying eqns (5.9) and (5.10) as they are not independent.

- Although the CD strength is a maximum when τ, α, γ are all $90°$, under these circumstances $V = 0$, so the positive and negative CD signals are centred at ε and exactly cancel, resulting in no CD signal.
- According to eqn (5.9) the CD strength should get larger as the distance between the chromophores increases. However, V concomitantly decreases so more and more cancellation occurs, thus avoiding the nonsense situation of two chromophores too far apart to interact having an infinitely large CD signal. The extent of cancellation depends upon the shapes of the bands but goes approximately linearly in V. From eqns (5.9) and (5.10) we thus conclude that the CD signal effectively decreases as R_{AC}^{-2}.
- The CD of two enantiomers may be seen to be equal and opposite by reflecting the system about the plane containing $\boldsymbol{\mu}^a$ and $\boldsymbol{R}_{AC}$, *i.e.* the x-z plane, as then $\tau \to 360° - \tau$, and $\sin \tau \to -\sin \tau$.

When $\tau = \alpha = \gamma = 90°$ the **A/C** system is achiral as it has an S_4 improper rotation axis.

5.5 Degenerate coupled-oscillator CD: some examples

Biphenyl: $\alpha = \gamma = 90°$

Eqns (5.9) and (5.10) are simplest when $\alpha = \gamma = 0°$ or $\alpha = \gamma = 90°$, *i.e.* when the transitions are polarized either parallel or perpendicular to the line connecting the chromophore origins. If we may ignore electron exchange (conjugation) between the rings, biphenyl (Fig. 5.4) in a twisted conformation provides us with such an example with each ring of biphenyl having two possible transition polarizations.

Consider first two degenerate transitions with $\alpha = \gamma = 0°$. The edtms for these transitions are co-planar (in fact co-linear), so they form an achiral system and have no *CD* strength. Consistent with this, substitution of $\alpha = \gamma = 0°$ into eqn (5.9) gives a value of 0 for the *CD* arising from this coupling.

If, however, two degenerate $\alpha = \gamma = 90°$ transitions couple then the resulting *CD* strengths and transition energies are

$$R^{\pm} = \pm \frac{\varepsilon \mu^2 R_{AC} \sin \tau}{4\hbar} \tag{5.11}$$

$$\varepsilon^{\pm} = \varepsilon \pm \frac{\mu^2 \cos \tau}{R_{AC}^3} \tag{5.12}$$

where τ is the dihedral angle traced in going in an anticlockwise direction when viewed down $\boldsymbol{R}_{AC}$ from $\boldsymbol{\mu}^a$ to $\boldsymbol{\mu}^c$ as illustrated in Fig. 5.7. When $\tau = 0°$ the *CD* strength is zero and when $\tau = 90°$ or $270°$ there is no energy splitting between the two bands, so no *CD* spectrum is observed. Maximum magnitude *CD* intensities are expected for τ in the region of $\pm 45°$ and $\pm 135°$ (though the actual value will depend upon the shapes of the bands). The gas phase experimental value for τ is $44°$ meaning each biphenyl molecule has close to the maximum possible *CD* signal.[4] However, in any sample there are equal proportions of both enantiomers, so we observe no experimental *CD* spectrum for biphenyl.

To understand the origin of the Fig. 5.7 spectra consider a short axis polarized transition moment in each phenyl oriented so that $\tau \approx 45°$. The two moments may couple in phase or out of phase. As may be seen from Fig. 5.7, $\boldsymbol{\mu}^c$ circles about $\boldsymbol{\mu}^a$ in a right-handed direction (put your right thumb along $\boldsymbol{\mu}^a$, and the right hand fingers may then be bent to lie along $\boldsymbol{\mu}^c$) as does $\boldsymbol{\mu}^a$ about $\boldsymbol{\mu}^c$. As the in-phase coupling of the two moments has net edtm $\boldsymbol{\mu} = \boldsymbol{\mu}^a + \boldsymbol{\mu}^c$, and the net mdtm of the in-phase coupled moments is a circling of charge about $\boldsymbol{\mu}$, the net effect of the in-phase coupling of the two moments is a right-handed helix of electron motion in space. Thus we expect a positive *CD* signal for $0 < \tau < 90°$ as predicted by eqn (5.11) and illustrated in Figs 5.5 and 5.7.

Before leaving our first example we note a few cautions in using *CD* to analyse molecular geometry.

- Despite the opposite handedness of the molecules giving the first and third spectra in Fig. 5.7, these spectra *look* the same because *both* the *CD* signs and the energy order are inverted in going from one to the other.

- In order to test experimentally Fig. 5.7 we would need to consider substituted biphenyl systems and it may not be appropriate to assume that the phenyl transitions are unchanged by the substitution. We shall therefore leave this example as simply an illustration of the simplest case of degenerate coupled-oscillator *CD*.

- The transitions used in the analysis must be *independent*, in other words there must not be significant *conjugation* between the sub-units into which we divide the molecule. For example, the π system of butadiene cannot satisfactorily be treated as two independent ethylene groups.

- Although we have stated that the coupling between transitions of identical chromophores is dominated by the degenerate coupling of one transition with the same one on the other chromophore, if other transitions are close in energy they may interfere significantly. For example, the coupling of $\pi \rightarrow \pi^*$ transitions with $\sigma \rightarrow \pi^*$ transitions seems to be the reason for the mixed success of assignments of the geometry of non-conjugated chiral dienes.[4]

The extent of conjugation is greater the more planar a molecule becomes, so if $\tau \sim 45°$ in biphenyl then we can assume there is little conjugation.

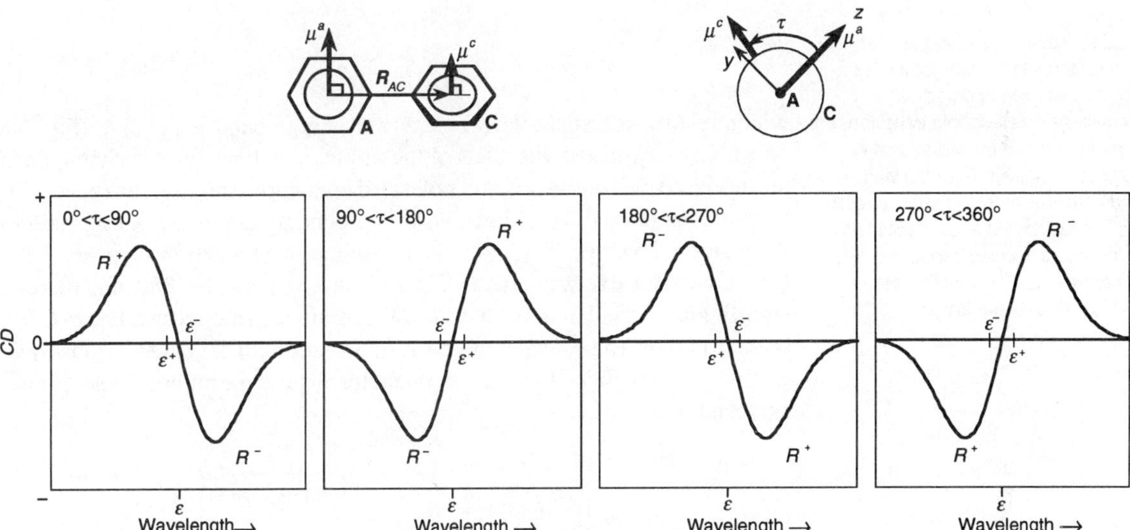

Fig. 5.7 Biphenyl and its exciton *CD* spectra for 'short axis' polarized transitions as a function of τ (assuming Gaussian band shapes for the two transitions).

6,15-Dihydro-6,15-ethanonaphthol[2,3-*c*]pentaphene: $\alpha = \gamma \neq 90°$

The next level of complexity is provided by the two anthracenes of the title molecule. Each anthracene chromophore has a weak short axis polarized transition at ~360 nm and a strong long axis one at 252 nm (Fig. 1.6). In contrast to the biphenyl case, the transition polarizations are neither parallel nor perpendicular to R_{AC} and τ is defined by the rigid geometry of the molecule (Fig. 5.8). For the long axis polarized transitions $\alpha^{||} = \gamma^{||} = 137°$ and $\tau^{||} = 259°$ for the enantiomer illustrated in Fig. 5.8 so eqns (5.9) and (5.10) lead us to expect a symmetric exciton couplet at ~260 nm with the lower energy (longer wavelength) band having the positive sign and the higher energy band the negative sign since

$$R^{\pm} = \pm \frac{\varepsilon \mu^2 R_{AC}}{4\hbar} \sin \alpha \sin \gamma \sin \tau$$

$$= \mp 0.46 \frac{\varepsilon \mu^2 R_{AC}}{4\hbar} \tag{5.13}$$

and the energies of the two bands are

$$\varepsilon^{\pm} = \varepsilon \pm \mu^2 \left(\sin \alpha \sin \gamma \cos \tau + 2 \cos \alpha \cos \gamma \right) R_{AC}^{-3}$$

$$= \varepsilon \pm 0.98 \mu^2 R_{AC}^{-3} \tag{5.14}$$

The angles for the short axis polarized transition are $\alpha^{\perp} = \gamma^{\perp} = 73°$ and $\tau^{\perp} = 112°$, so

$$R^{\pm} = \pm 0.85 \frac{\varepsilon \mu^2 R_{AC}}{4\hbar} \tag{5.15}$$

and

$$\varepsilon^{\pm} = \varepsilon \mp 0.17 \mu^2 R_{AC}^{-3} \tag{5.16}$$

which results in a spectrum whose lower energy band is positive (Fig. 5.8). As might be expected, the signs of the in-phase and out-of-phase components are reversed for the short axis polarized transitions relative to those of the long-axis polarized transitions, however, concomitantly the energy order of the bands is reversed. Therefore the *CD* of the two transitions appears to have the same sign exciton effect. The *CD* strength in the 380 nm region is significantly smaller than that at 260 nm since coupled-oscillator *CD*, as seen from eqn (5.9), scales with oscillator strength μ^2 (normal absorption intensity) as well as the geometry factor. The experimental spectrum[6] is sketched in Fig. 5.8.

An additional feature of the 380 nm band is its unusual sign oscillations. When performing coupled-oscillator calculations on electronic transitions we usually ignore the vibrational structure. However, here the vibronic bands are well resolved and we really should consider the exciton coupling of the individual vibronic bands and plot the net overlap. The result of this is precisely what is observed.

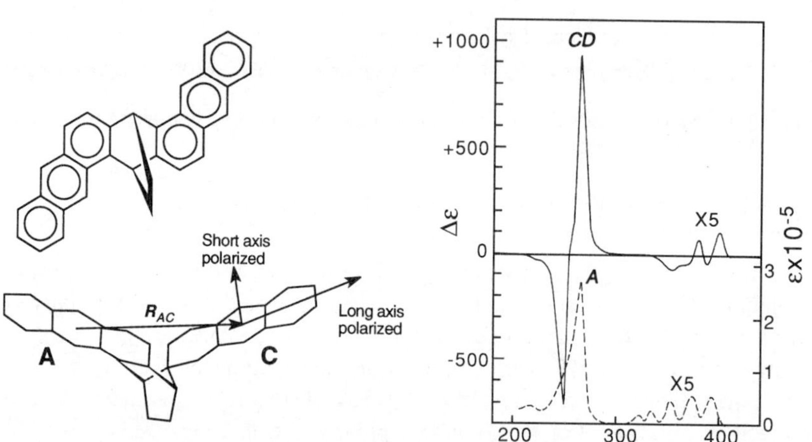

Fig. 5.8 Geometry, absorption spectrum, and *CD* spectrum of 6*R*,15*R*-(+)-dihydro-6,15-ethanonaphthol[2,3-*c*]pentaphene.

Rather than simply show that we can write equations to reproduce the experimental *CD* spectra, the usual application would be to determine which enantiomer we had in a sample by finding the value of τ that gives the experimentally observed *CD* signs. If more detailed geometric information is required of course the *CD* magnitudes also become important in the analysis.

In-ligand transition of a bis-chelate metal complex: $\alpha \neq \gamma \neq 90°$

Although metal complexes[7] are chemically different from the previous example, the *CD* arising from the coupling of long axis polarized transitions of the ligands in bis-chelate transition metal complexes can be understood in the same way as for the bis-anthracene molecule. In this application the metal is not part of any chromophore, but can be considered as a structural unit placing the ligands in a chiral orientation in space relative to each other. For simplicity, let us assume the ligating atoms define a perfect octahedron, then a Δ enantiomer has the coordinates and dipoles shown in Fig. 5.9.

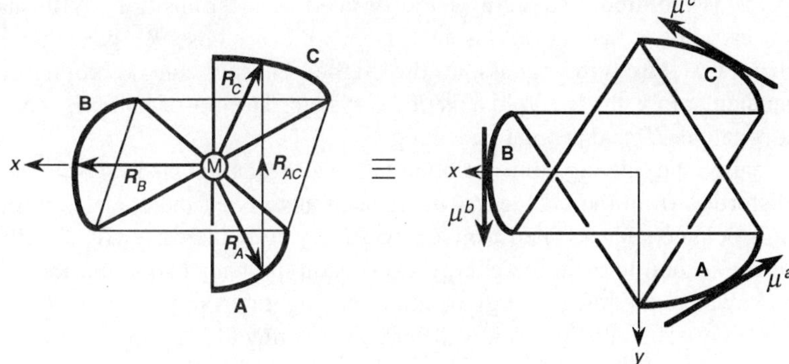

Fig. 5.9 Geometry described in the text for Δ bis and tris chelate complexes.

The axis system used in Fig. 5.9 is that defined by the $\mathbf{D_3}$ symmetry of the next example rather than that of Fig. 5.6. The vectors from the metal to the centres of the chelates $\mathbf{A}$ and $\mathbf{C}$ are:

$$R_A = \rho\left(-\frac{1}{2}, \frac{\sqrt{3}}{2}, 0\right) \tag{5.17}$$

and

$$R_C = \rho\left(-\frac{1}{2}, -\frac{\sqrt{3}}{2}, 0\right) \tag{5.18}$$

so

$$R_{AC} = \sqrt{3}\rho(0, -1, 0) \tag{5.19}$$

where ρ is the distance from the metal to the centre of a chelate. The edtms as illustrated are:

$$\mu^a = \mu\left(-\frac{1}{2}, -\frac{1}{2\sqrt{3}}, \frac{\sqrt{2}}{\sqrt{3}}\right) \tag{5.20}$$

and

$$\boldsymbol{\mu}^c = \mu\left(\frac{1}{2}, -\frac{1}{2\sqrt{3}}, \frac{\sqrt{2}}{\sqrt{3}}\right) \tag{5.21}$$

from which it follows (Fig. 5.6) that $\cos\alpha = -\cos\gamma = (2\sqrt{3})^{-1}$, $\sin\alpha = \sin\gamma = \sqrt{(11/12)}$ and $\cos\tau = 5/11$. Further, CAR is righthanded so $0 < \tau < 180°$ from which it follows that $\sin\tau = 4\sqrt{6}/11$.

The *CD* strengths and transition energies for the *CD* arising from the coupling of long axis polarized transition moments on two chelates is thus [from eqns (5.6) and (5.8) or eqns (5.9) and (5.10)]

$$R^{\pm} = \pm\frac{\varepsilon\mu^2\rho}{2\sqrt{2}\hbar} \tag{5.22}$$

$$\varepsilon^{\pm} = \varepsilon \pm \frac{\mu^2}{12\sqrt{3}\rho^3} \tag{5.23}$$

where μ^2 is the dipole strength of the isolated ligand transition. With the directions chosen for the edtms in Fig. 5.9 the in-phase R^+ band has a positive *CD* strength, occurs at the higher energy, and is polarized perpendicular to x, the two-fold axis of the system. The out-of-phase R^- band has a negative *CD* and is polarized along x.

In reality, the ligating atoms seldom define a regular octahedron. Imagine the distortion from the octahedral being such as to twist the chelates more towards being coplanar. Then τ increases. A *very* large increase will actually cause the second term in the energy expression to change sign and the R^+ state to swap to the lower energy position making it appear that the *CD* sign has inverted—although in fact it is the energy ordering that has inverted.

In-ligand transition of a tris-chelate metal complex: $\alpha_1 = \alpha_2 = \alpha_3 \neq 90°$

The previous example leads directly to the in-ligand *CD* for Δ tris-chelate transition metal complexes.[7] As there are now three chelates we also require vectors for chelate **B** in Fig. 5.9:

$$R_B = \rho(1, 0, 0) \tag{5.24}$$

$$\boldsymbol{\mu}^b = \mu\left(0, \frac{1}{\sqrt{3}}, \frac{\sqrt{2}}{\sqrt{3}}\right) \tag{5.25}$$

to evaluate the *CD*. However, since the system is symmetric, the coupling of each pair of ligands leads to the same contribution, so we can simply multiply eqn (5.22) by a factor of 3. In fact, a further symmetry factor of $\frac{2}{3}$ is required (see Chapter 7) to account for the higher symmetry of the system.

The *CD* strengths and transition energies for the in-ligand *CD* of a Δ configuration tris-chelate are then:

$$R\left(\frac{z}{x/y}\right) = R\left(\frac{A_2}{E}\right) = \pm\frac{\varepsilon\mu^2\rho}{\sqrt{2}\hbar} \tag{5.26}$$

$$\varepsilon\left(\frac{z}{x/y}\right) = \varepsilon\left(\frac{A_2}{E}\right) = \varepsilon \pm \frac{\mu^2}{12\sqrt{3}\rho^3} \tag{5.27}$$

$R(z)$ is the z-polarized *CD* band (A_2) that results from the in-phase coupling of the three dipoles. It occurs at the higher energy and is positive. The out-of-phase couplings give the second *CD* band which has the same magnitude but opposite sign from the first band. Its polarization is x/y or E—its transition density is spread out over the whole x/y plane. An experimental spectrum is illustrated in Fig. 5.10 (for the mirror configuration Λ).

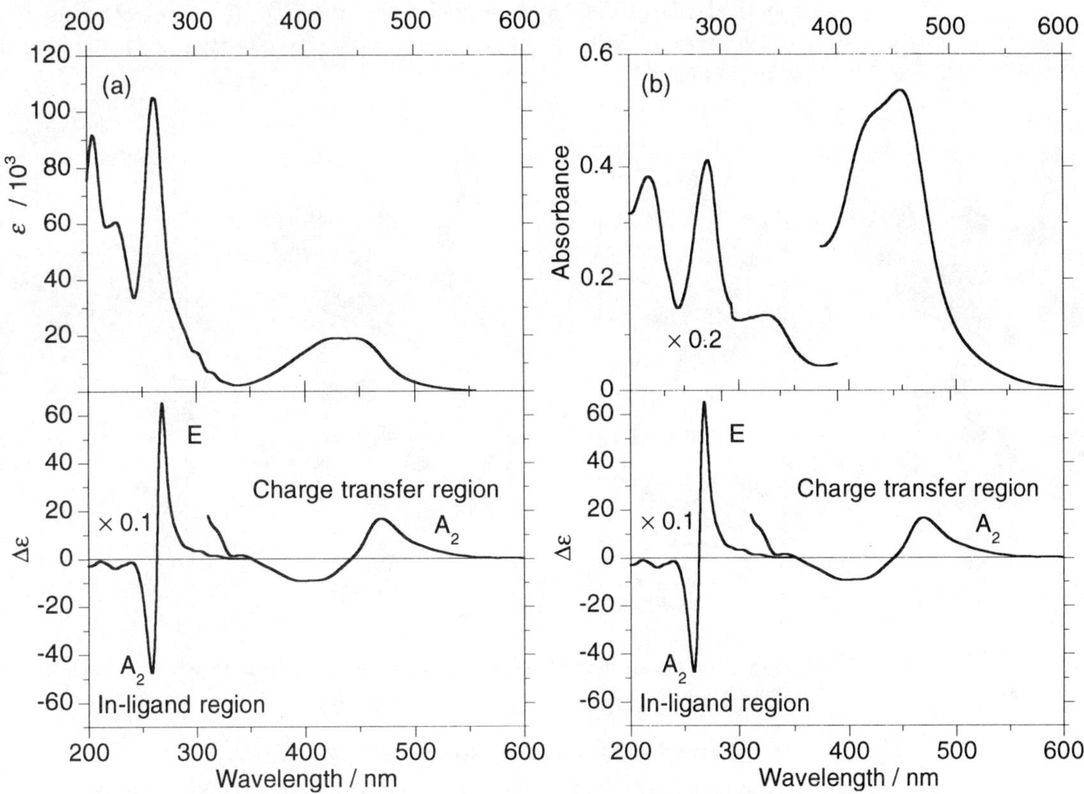

Fig. 5.10 Absorbance and *CD* spectrum of (a) Λ-[Ru(4,7-dimethyl-1,10-phenanthroline)$_3$]$^{2+}$ and (b) Λ-{Ru[(2,2'-bipyridine)$_2$(pyridine)$_2$]$^{2+}$ (0.07 mg /cm^3). (See Fig. 6.10 for chelate structures.)

Helical peptide

A somewhat simplistic model for determining the backbone $\pi \rightarrow \pi^*$ *CD* of α-helical peptides is to assume we may represent the peptide by a right-handed helix of transition moments oriented along each C=O bond (Fig. 2.4) as illustrated in Fig. 5.11 (*cf.* §3.2). Each neighbouring pair of dipoles has $\tau \sim 300°$ if they are both directed towards the oxygen, and $\tau = 120°$ when they have opposite orientations. Although N residues would lead to N transitions from the couplings, we need only consider the highest and lowest energy ones; the others will essentially cancel each other out between the two extrema.

The lowest energy excited state (and hence longer wavelength component of the transition) will result from the head-to-tail couplings of neighbouring transition moments. For an α-helix, head-to-tail coupling of neighbouring $\pi \to \pi^*$ edtms occurs when they are both pointing from C to O along the C=O bond. As Fig. 5.11 illustrates, the net edtm of this state is parallel to the helix axis, and the net mdtm is antiparallel to it, making a net left-handed helix of electron displacement and therefore a negative *CD*. The head-to-head, tail-to-tail coupling will give the highest energy transition whose polarization is perpendicular to the helix axis and has a net right-handed helical charge displacement and a positive *CD*. The two *CD* components (190 nm and 208 nm) of the $\pi \to \pi^*$ α-helix *CD* (Fig. 2.5) illustrate this discussion.

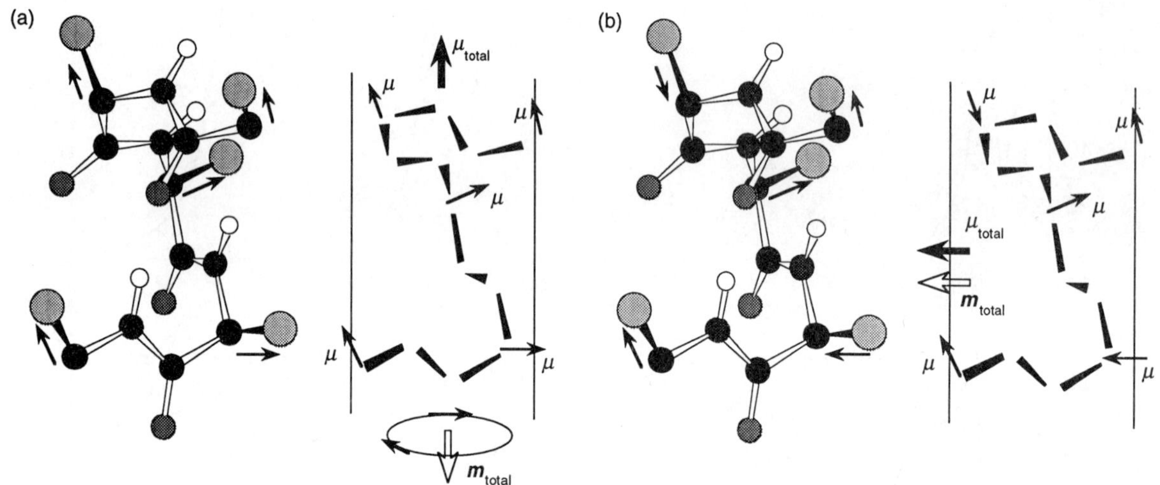

Fig. 5.11 Model of an α-helical peptide indicating (a) head-to-tail and (b) head-to-head couplings of the edtms.

Combined *CD* and *LD* spectroscopy

The coupled $\pi \to \pi^*$ transitions of the amide chromophores in polypeptides provide an excellent example of how *CD* and *LD* spectroscopy can give complementary structural information. As noticed above, the two coupling modes that give rise to the high energy positive *CD* and the low energy negative *CD* bands are associated with net electric transition moments that lie, respectively, perpendicular and parallel to the helix axis and thus give rise to respectively negative and positive *LD* bands in, for example, a flow *LD* spectrum. Since the net μ is simply a vector sum of the μs of the individual chromophores (with appropriate signs to match the phase of the coupling mode) the *LD* signal directly relates to the structure of the polypeptide and may be analysed as outlined in Chapter 3.

5.6 Non-degenerate coupled-oscillator *CD*: general case

In the final sections of this chapter we shall examine the *CD* spectra we could expect to observe if we have two non-identical achiral chromophores **A** and **C** whose eda transitions couple. The required induced mdtms arise in the same way as when degenerate oscillators couple—a linear motion of charge at a distance (in another chromophore) has a circular (magnetic) component locally. The main differences between the non-degenerate and degenerate cases are:

- the two *CD* bands resulting from the coupling of two non-degenerate transitions occur at (strictly very close to) the energies of the two non-degenerate transitions, rather than as an exciton couplet where most of the *CD* intensity is cancelled, and
- the coupling between non-degenerate transitions is not as strong as that between degenerate transitions and it *decreases* when either the energy difference between the transitions or the distance between the chromophores *increases*.

The Fig. 5.6 coordinate system used for the degenerate coupled-oscillator analysis can be used for the non-degenerate situation. The only change in notation required is that we need to distinguish the transition energies and transition moment magnitudes of the transitions in the two chromophores since they are now different. As shown in Chapter 7, the *CD* induced at ε_a into the transition on **A** by its coupling with an edtm of different energy in **C** is

$$R(\varepsilon_a) = \frac{-\varepsilon_a \varepsilon_c V}{\hbar\left(\varepsilon_c^2 - \varepsilon_a^2\right)} \left\{ \boldsymbol{\mu}^c \times \boldsymbol{\mu}^a \cdot \mathbf{R}_{AC} \right\}$$

$$= \frac{-\varepsilon_a \varepsilon_c}{\hbar\left(\varepsilon_c^2 - \varepsilon_a^2\right)} \left\{ \frac{\boldsymbol{\mu}^c \cdot \boldsymbol{\mu}^a - 3\hat{\mathbf{R}}_{AC} \cdot \boldsymbol{\mu}^a \boldsymbol{\mu}^c \cdot \hat{\mathbf{R}}_{AC}}{R_{AC}^2} \right\} \left\{ \boldsymbol{\mu}^c \times \boldsymbol{\mu}^a \cdot \hat{\mathbf{R}}_{AC} \right\}$$

$$(5.28)$$

The *CD* induced at ε_c is determined by simply permuting all the '*a*' and '*c*' and '*A*' and '*C*' labels in eqn (5.28). Thereby the term $\left(\varepsilon_c^2 - \varepsilon_a^2\right)$ in the denominator changes sign so $R(\varepsilon_c) = -R(\varepsilon_a)$

The alternative version of eqn (5.28) in terms of the angles defined in Fig. 5.6 is given below. As in the degenerate coupled-oscillator case, it is very important to define the sense of τ, which determines the chirality of the system correctly.

$$R(\varepsilon_a) = \frac{-\varepsilon_a \varepsilon_c (\mu^a \mu^c)^2}{\hbar\left(\varepsilon_c^2 - \varepsilon_a^2\right) R_{AC}^2} \left\{ \sin\alpha \sin\gamma \cos\tau + 2\cos\alpha \cos\gamma \right\} \sin\alpha \sin\gamma \sin\tau$$

$$(5.29)$$

A number of general points about non-degenerate coupled-oscillator *CD* spectra follow from eqns (5.28) and (5.29).

- The *CD* strength *decreases* with *increasing* **A** to **C** distance according to R_{AC}^2. This is the smallest distance-dependence for any *CD* mechanism; coupled-oscillator *CD*s are therefore generally larger than ones arising from any of the chromophore coupling mechanisms discussed in the next chapter (hence if a transition is eda and mda but the electric and magnetic moments are perpendicular, then the coupled-oscillator mechanism generally dominates the observed *CD*).
- $R^{\pm} \to -R^{\pm}$, *i.e.* $CD \to -CD$, for the mirror image system.

- It is often the case that $\alpha = \gamma = 90°$. When this situation arises the geometry factor in eqn (5.29) is

$$\cos\tau\sin\tau = \frac{1}{2}\sin(2\tau) \qquad (5.30)$$

so the *CD* is a maximum when $\boldsymbol{\mu}^a$ and $\boldsymbol{\mu}^c$ are oriented at 45° to one another.

- If we wish to know the combined effect of many transitions in **C** on an **A** transition, we simply introduce a sum over c into eqns (5.28) and (5.29).

- The *CD* induced into the ε_c transition of **C** by $\boldsymbol{\mu}^a$ follows upon permuting the 'a' and 'c' labels and the 'A' and 'C' labels in eqns (5.28) and (5.29) (note $R_{AC} = -R_{CA}$). Thus, the *CD* induced by an **A** transition into a **C** transition is equal in magnitude and opposite in sign from the one induced by the **C** transition into the **A** transition as illustrated in Fig. 5.12.

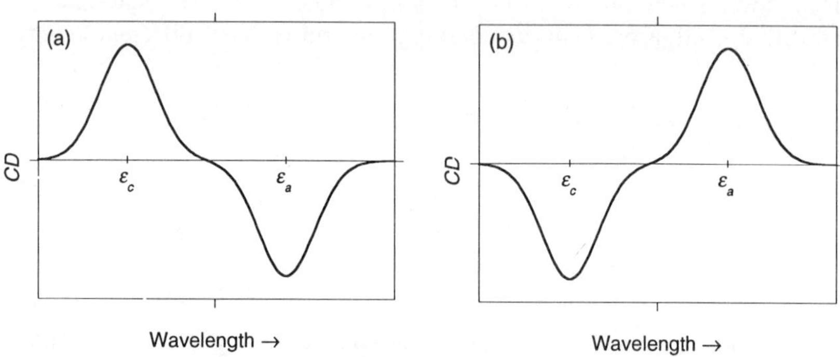

Fig. 5.12 Schematic illustration of non-degenerate coupled-oscillator *CD* spectra for $\alpha = \gamma = 90°$: (a) $0 \leq 2\tau < 180°$ and (b) $180° \leq 2\tau < 360°$ (see Fig. 5.6 for defintion of τ).

5.7 Non-degenerate coupled-oscillator *CD*: some examples

DNA dinucleotide: $\alpha = \gamma = 90°$

Consider two non-identical DNA nucleo-bases (*e.g.* guanine and adenine) that are a small fragment of B-DNA, so they are vertically stacked parallel (but skew) to one another with their planes perpendicular to the vector between their origins (Fig. 5.13). Thus, $\alpha = \gamma = 90°$. With ten bases per turn of DNA, the twist between the long-axes of adjacent base pairs is 36°. For simplicity we shall assume that each base has only one transition moment polarized either parallel to its long axis or perpendicular to it. We denote the base that has the lower energy transition **A**, and look at the dinucleotide from above with **A** on top. Fig. 5.13 shows the value of τ for the possible transition moment polarization combinations of this simple model.

The *CD* induced into the transition of the top base may be determined from eqn (5.28) or (5.29)

Although this is a fairly unrealistic description of DNA it forms a starting point for fuller theoretical treatments.[10–12]

$$R(\varepsilon_a) = \frac{-\varepsilon_a\varepsilon_c\left(\mu^a\mu^c\right)^2}{2\hbar\left(\varepsilon_c^2 - \varepsilon_a^2\right)R_{AC}^2}\sin(2\tau) \qquad (5.31)$$

so the 'long–long' and 'short–short' combinations both give a positive *CD* at ε_a, and the mixed combinations both give a negative *CD*. The *CD* signals at ε_c are equal in magnitude but opposite in sign from those at ε_a (Fig. 5.13).

That the ε_c *CD* strengths are opposite in sign from those at ε_a may be seen by redrawing Fig. 5.13 so that C is on top. Then exchange the **A** and **C** labels in the equations. The 'new' value for τ is the same as the old one, but the energy denominator has become negative.

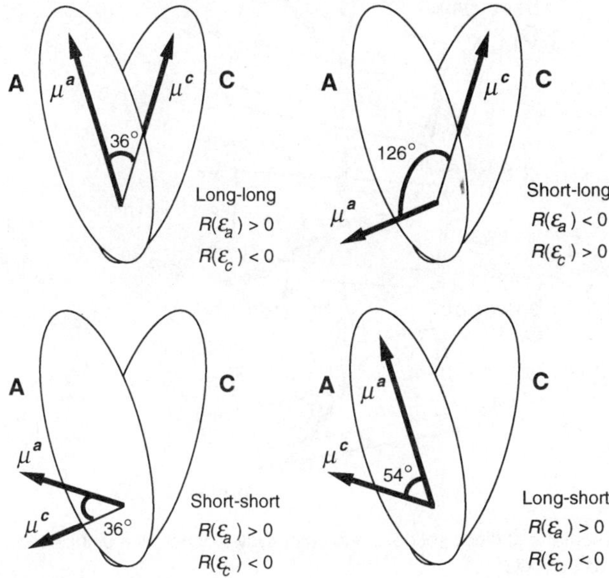

Fig. 5.13 Geometry and *CD* signs (assuming $\varepsilon_a < \varepsilon_c$) for a 'B-DNA' dinucleotide with two different bases. Each nucleotide is assumed to have only one edtm, either long axis or short axis polarized.

It seems immediately reasonable from the above discussion that left-handed Z-DNA should have a *CD* signal more or less a mirror image of that found for B-DNA (*cf.* Fig. 2.10) on the basis of eqn (5.31). While this might be approximately true, we must realize that in reality the base transition moments are not only long or short axis polarized and Z-DNA is not a simple antisymmetric opposite twist of the DNA bases from that found in B-DNA since the presence of the chiral ribose moieties makes it a diastereomeric system. A further warning about the complexities of reality arises when one considers that each base has several transitions occurring at similar transition energies and DNA has four different bases. Experimental DNA spectra are thus the result of complicated coupling patterns between many overlapping transitions.

For the case of two identical bases in a B-DNA conformation (a degenerate coupled-oscillator example),
$$\tau = (360 - 36)^\circ$$
R^+, occurring at the higher energy, will be negative and R^- at lower energy positive, explaining the common exciton-like pattern of the long wavelength region of B-DNA spectra (*e.g.* Fig. 2.9).

Intercalator in a random sequence DNA polynucleotide

In this example we wish to determine the *CD* induced into a transition of an achiral planar aromatic molecule that is intercalated (sandwiched) between two base pairs of a B-DNA molecule. The plane of the intercalator lies parallel to those of the bases, so as in the previous example $\alpha = \gamma = 90°$. Thus eqn (5.31), where $\boldsymbol{\mu}^a$ is now the intercalator transition moment, describes the *CD* for this situation. A sum over at least the neighbouring base pairs and all their transitions should also be included. The geometry of the dinucleotide / adduct situation is illustrated in Fig. 5.14.

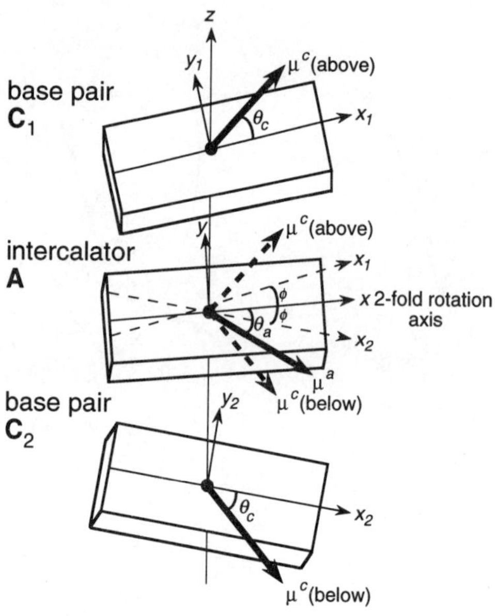

Fig. 5.14 Geometry of dinucleotide/ intercalator system where the dinucleotide has a two-fold rotation axis about x.

In a totally random sequence DNA with random intercalative binding on average we can expect to find a particular base pair and its two-fold rotation (5'-T–A-3' and 5'-T–A-3', say) equally often about a given intercalator. From Fig. 5.14 τ (above) = $(360 - \theta_c - \theta_a - \phi)°$ and τ (below) = $(360 - \theta_c + \theta_a - \phi)°$. The *CD* from the coupling of the intercalator transition with the one transition on each base is, in terms of the angles defined in Fig. 5.14, therefore [*cf.* eqn (5.29)]

$$R(\varepsilon_a) = \frac{\varepsilon_a \varepsilon_c \left(\mu^a \mu^c\right)^2}{2\hbar\left(\varepsilon_c^2 - \varepsilon_a^2\right)R_{AC}^2}\left\{\sin 2(\theta_c - \theta_a + \phi) + \sin 2(\theta_c + \theta_a + \phi\right\}$$

$$= \varepsilon_a\left(\mu^a\right)^2 \cos(2\theta_a)\left\{\frac{\varepsilon_c\left(\mu^c\right)^2 \sin(2\theta_c + 2\phi)}{2\hbar\left(\varepsilon_c^2 - \varepsilon_a^2\right)R_{AC}^2}\right\}$$

$$(5.32)$$

The $\cos(2\theta) = (2\cos^2\theta - 1)$ dependence makes the *ICD* behave as a 'microscopic *LD*', not depending on the sign of θ.

The sign of $\sin(2\theta_c + 2\phi)$ in eqn (5.32) is determined by the DNA helix pitch (2ϕ in Fig. 5.14) and the polarizations of transition moments within the base pairs [recall a summation over all base transitions is implied in eqn (5.32)]. The sign of $\cos(2\theta_a)$ is determined by the polarization of the intercalator transition and the orientation of the intercalator in the pocket. Schipper *et al.*[13] deduced that the average sign of the DNA factor (the term in parentheses) is negative. Therefore, long axis polarized transitions of intercalators oriented with their long axis 'parallel' to the DNA bases (thus having $\theta_2 \sim 0$) will have a negative induced *CD* signal in a random sequence DNA. Ones whose long axis pokes out into the DNA grooves will have a positive induced *CD* signal for long axis polarized transitions. The converse signs are to be expected for short axis polarized transitions.

As must be the case, the same geometry/*CD* correlations are derived for a random sequence DNA if y is used as the two-fold rotation axis rather than x. The symmetry-related base pairs would then be *e.g.* 5'-T–A-3' and 5'-A–T-3'. However, it should be noted that eqn (5.32) does not apply directly to alternating homopolymers such as poly[d(G–C)]$_2$ since the value of θ_c is not the same for the same transition in 5'-G–C-3' and in 5'-C–G-3', but is a function of the helix pitch. Thus, the *CD* induced into an intercalator in a purine–pyrimidine site is expected to be quite different from when it is in a pyrimidine–purine site. In fact, calculations have shown that the sign of the induced signal can be a function of site.[12]

Eqn (5.32), though not necessarily the sign of the DNA factor, also holds for B-DNA non-alternating homopolymers such as [poly(dG)].[poly(dC)].

Charge transfer transitions of tris-chelate metal complexes: $\alpha = \gamma = 90°$

In addition to eda in-ligand transitions, tris-chelate transition metal complexes also have eda charge transfer transitions, where electron density is transferred from metal to ligand orbitals or *vice versa*. We define the charge transfer chromophore, **A**, to be the metal and the directly ligating atoms and take it to have $\mathbf{D}_{3d}$ symmetry so it may be treated as an achiral guest in the middle of the chiral host composed of the three chelates $\mathbf{C}_1$, $\mathbf{C}_2$, and $\mathbf{C}_3$ (Fig. 5.15). **A** transitions are polarized either along its three-fold axis, z, for which $\alpha = 90°$, or anywhere in the plane perpendicular to z. The two-fold degenerate x/y polarized transitions may be considered as one polarized along x (*i.e.* pointing towards a chelate) with $\alpha = 0°$, so no *CD*, and one polarized along y with $\alpha = 90°$ as illustrated in Fig. 5.15. For planar aromatic chelates we need only consider the long axis polarized transitions (short axis polarized ones, which are co-planar, give no *CD*) we used for calculating the in-ligand *CD*, thus $\gamma = 90°$ for each $\mathbf{C}_i$ transition.

The *CD* for a z polarized (along the three-fold axis of the molecule) charge transfer transition of a Δ complex as drawn in Fig. 5.15 has $\cos \tau = \sqrt{2}/\sqrt{3}$ and $\sin \tau = 1/\sqrt{3}$, so from eqn (5.29)

$$R(z) = R(A_2)$$

Note that $(\mu^a)^2$ is the oscillator strength for the A_2 charge transfer band; it has half the intensity of the E band. $(\mu^c)^2$ is the oscillator strength of the in-ligand transition.

$$= \frac{-3\varepsilon_a\varepsilon_c\mu^a\mu^c}{\hbar\left(\varepsilon_c^2 - \varepsilon_a^2\right)R_{AC}^2}\sin^2\alpha\sin^2\gamma\cos\tau\sin\tau \qquad (5.33)$$

$$= \frac{-\sqrt{2}\varepsilon_a\varepsilon_c\mu^a\mu^c}{\hbar\left(\varepsilon_c^2 - \varepsilon_a^2\right)R_{AC}^2}$$

where the factor of three comes from the three chelates, and the A_2 label is the symmetry label for the transition. Note, this is opposite in sign from the same polarization in-ligand *CD* band and also equal in magnitude but opposite in sign from the *x/y* charge transfer band since $\tau(E) = 360° - \tau(A_2)$

$$R(A_2) = -R(E) \tag{5.34}$$

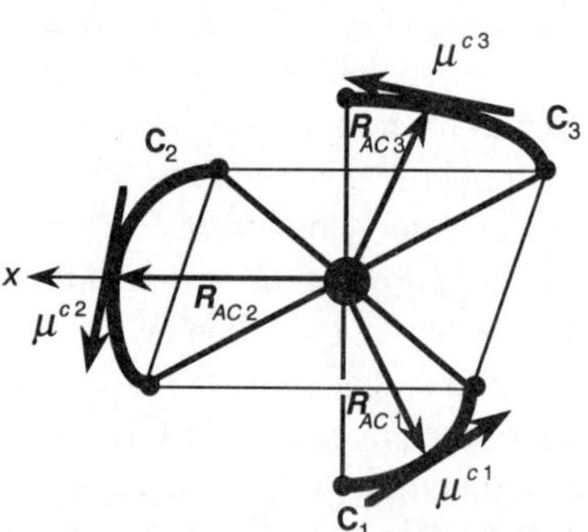

Fig. 5.15 Geometry of the tris-chelate charge transfer system. Components of the **A** chromophore are indicated by solid circles and the three **C** chromophores by thick curves. Axis system as in Fig. 5.9.

The energy ordering of the charge transfer transitions is difficult to determine, in contrast to the in-ligand case. As the charge transfer and in-ligand *CD* have *opposite* signs for a given polarization, the polarizations of the charge transfer bands can thus be determined from the *CD* spectrum.

The resulting *CD* spectrum (Fig. 5.10) at the charge transfer transition energy looks similar to the in-ligand region spectrum. In the latter case the overlapping bands are the result of two degenerate transitions on *different* chelates coupling together to give two transitions that are close in energy with opposite sign *CD*s. In the charge transfer case, however, two nearly degenerate (the A_2 and E) charge transfer transitions on the *same* chromophore each non-degenerately couple with chelate transitions. The two *CD* bands have opposite signs because the A_2 and E transitions are perpendicularly polarized.

In-ligand transitions of unsymmetric bis-chelate metal complexes: $\alpha = \gamma \neq 90°$

In the previous section we determined the *CD* resulting from the coupling of in-ligand long-axis polarized transition moments on two identical chelates in a *bis*-chelate transition metal complex. If we now consider two distinct ligands (such as 1,10-phenanthroline and 2,2'-bipyridine) we can use the same geometry as in the previous bis-chelate example, but must use the non-degenerate coupled-oscillator equation, eqn (5.29). Thus, the *CD* expected at ε_a is (see eqn (5.17) for notation):

$$R(\varepsilon_a) = \frac{-\varepsilon_a\varepsilon_c\left(\mu^a\mu^c\right)^2}{2\sqrt{6}\hbar\left(\varepsilon_c^2 - \varepsilon_a^2\right)R_{AC}^2}$$

$$= \frac{-\varepsilon_a\varepsilon_c\left(\mu^a\mu^c\right)^2}{6\sqrt{6}\hbar\left(\varepsilon_c^2 - \varepsilon_a^2\right)\rho^2} \tag{5.35}$$

The ε_c transition has *CD* equal in magnitude but opposite in sign from that of the transition at energy ε_a. The exact 'equal and opposite' is in fact seldom observed because in practice more than one transition on each chromophore takes part in the coupling. Experimental data may be found in *e.g.* reference 14.

Cyclodextrin inclusion compounds

The *CD* of large formally rather complicated systems is sometimes as simple as that of much smaller ones, especially if the system has high symmetry. One such system is a cyclodextrin guest/host complex (Fig. 5.16). Cyclodextrins are α–1,4-linked D-glucose oligomers. α-cyclodextrin has six glucose units, β-cyclodextrin seven, and so on.

Since, cyclodextrins have been found able to catalyse reactions involving their guests, it is of interest to know the orientation of the guest inside the cavity. The *CD* induced into the transitions of known polarization of the guest may tell us that immediately. Most *CD* experiments have been done with β-cyclodextrin since its cavity has just the right size to accommodate a single aromatic molecule such as benzene or naphthalene.

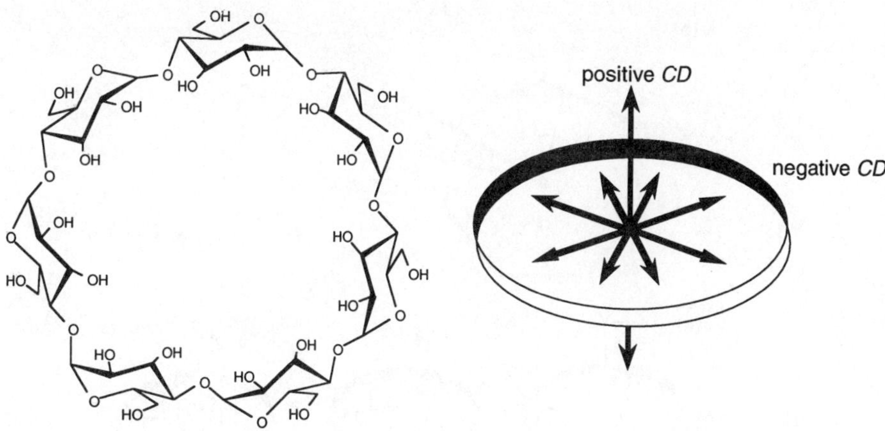

Fig. 5.16 β-cyclodextrin (left) and the signs of the *CD* induced into guest transitions of different polarizations within the cyclodextrin cavity (right). Arrows indicate transition polarizations.

We shall make two physically reasonable assumptions about the geometry of the host/guest system.

- One of the symmetry-determined axes of the guest molecule (chromophore **A**) aligns on average with the seven-fold rotation axis of β-cyclodextrin. This means that any guest transition is either polarized along z, the symmetry axis of the cyclodextrin, so $\alpha = 90°$, or is polarized perpendicular to z. In the latter case (see Fig. 5.17) if $\alpha = \theta$ for the coupling with glucose number 1, then $\alpha = \theta + (360 / 7)°$ for the second, and so on. A number of different transitions on each glucose will be important and their γ and τ values will all be different.

- The guest lies at the centre of the cyclodextrin, so

$$R_{AC} = \left(\cos\left(\frac{(n-1)360°}{7} \right), \sin\left(\frac{(n-1)360°}{7} \right), 0 \right) \qquad (5.36)$$

where n indexes the glucose units.

Let us consider the case of guests with non-degenerate transitions. The *CD* induced into a z-polarized transition by each glucose unit is the same, so the net *CD* induced by a β-cyclodextrin is seven times that of glucose number one, which is located with its origin along the x-axis. For a z-polarized guest transition eqn (5.29) thus becomes

$$R(\varepsilon_a, z) = \frac{-7\varepsilon_a \left(\mu^a \right)^2}{\hbar R_{AC}^2} \left\{ \sum_c \frac{\varepsilon_c \left(\mu^c \right)^2 \sin^2 \gamma \sin \tau \cos \tau}{\left(\varepsilon_c^2 - \varepsilon_a^2 \right)} \right\} \qquad (5.37)$$

where the factor of '7' accounts for the seven glucose units (*cf.* Table 7.1) and the summation is over all transitions on glucose unit number one.

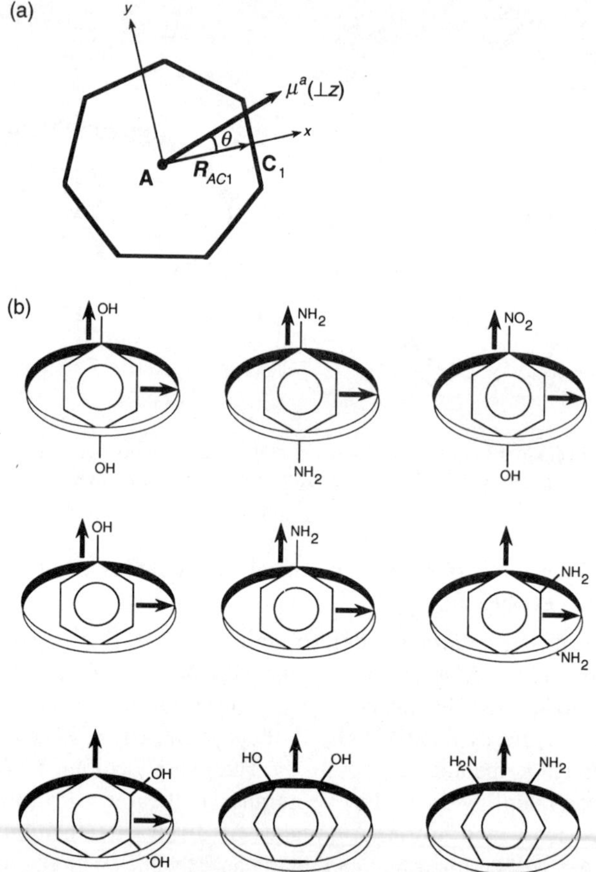

Fig. 5.17 (a) Geometric parameters for a guest transition polarized perpendicular to *z*. (b) Some cyclodextrin–guest complexes showing the guest transition polarizations (thick arrow), the sign of the induced *CD*, and the orientation of the guest consistent with these data.

If the guest lies oriented in the cavity so that its transition moment is in the x/y plane then the induced *CD* is half the magnitude of eqn (5.37) and of opposite sign (*cf.* Table 7.1). The term in large brackets in eqn (5.37) may be treated as an integral parameter of a specific glucose unit since for typical guest molecules (which are aromatic) $\varepsilon_c >> \varepsilon_a$, so $\varepsilon_c^2 - \varepsilon_a^2 \approx \varepsilon_c^2$. Therefore to determine the orientation of a guest within the cyclodextrin cavity we need only to know the sign of the integral glucose parameter. This can be deduced empirically from the sign of the dextrin-induced *CD* of *para* disubstituted benzene molecules which will have their long axis lying along z for steric reasons. Some results are summarized in Figs 5.16 and 5.17.[9]

The reduction of the *CD* intensity to 1/2 of the z-polarized value if a guest transition is polarized in the x/y plane is the result of the averaging due to the symmetry of the cyclodextrin host.

References

(1)   Newbury, S. F.; McClellan, J. A.; Rodger, A. *Anal. Commun.* **1996**, *33*, 117

(2)   Allenmark, S.; Oxelbark, J. *Enantiomer* **1996**, *1*, 13

(3)   Lightner, D. A.; Bouman, T. D.; Wijekoon, W. M. D.; Hansen, A. *J. Amer. Chem. Soc.* **1985**, *89*, 5805

(4)   Rubio, M.; Merchan, M.; Ortí, E.; Roos, B. O. *Chem. Phys. Lett.* **1995**, *234*, 373.

(5)   Wiegang, O. E. Jr. *J. Amer. Chem. Soc.* **1979**, *101*, 1965

(6)   Harada, N.; Takuma, Y.; Uda, H. *J. Amer. Chem. Soc.* **1976**, *98*, 5408

(7)   Bosnich, B. *Acc. Chem. Res.* **1969**, *2*, 266

(8)   Coggan, D. Z.; Bates, P. J.; Rodger, A. *Unpublished results*

(9)   Schipper, P. E.; Rodger, A. *J. Amer. Chem. Soc.* **1983**, *105*, 4541

(10)   Rizzo, V.; Schellman, J. A. *Biopolymers* **1984**, *23*, 435

(11)   Tinoco, I. J. *Adv. Chem Phys.* **1962**, *4*, 113

(12)   Lyng, R.; Rodger, A.; Nordén, B. *Biopolymers* **1992**, *31*, 1709; *32*, 1201

(13)   Schipper, P. E.; Nordén, B.; Tjerneld, F. *Chem. Phys. Lett.* **1980**, *70*, 17

(14)   Hidaka, J.; Douglas, B. E. *Inorg. Chem.* **1964**, *3*, 1180

6 Analysis of circular dichroism: magnetic dipole allowed transitions and magnetic *CD*

6.1 Introduction

Most of this chapter is devoted to the *CD* of magnetic dipole allowed (mda)/ electric dipole forbidden (edf) transitions. In the final section we shall also briefly examine the *CD* induced into electric dipole allowed (eda)/magnetic dipole forbidden (mdf) transitions by a permanent magnetic field: the so-called magnetic *CD* or *MCD*. The mechanisms giving rise to the *CD* of mda transitions and to *MCD* are quite different, but they naturally fit together since, in addition to their similar titles, both techniques often enable transitions whose existence is undetected or merely suspected in a normal absorption spectrum to be identified.

6.2 Magnetic dipole allowed transitions

In §5.2 we defined a magnetic dipole transition moment (mdtm) to be an axial vector, $\boldsymbol{m}$, which interacts with the magnetic field of the electromagnetic radiation to cause a circling of electron density about the transition polarization direction (Fig. 6.1). An example of a transition which, to a first approximation, has an mdtm but no edtm is the $n \to \pi^*$ transition of carbonyls that is illustrated in Fig. 6.1b.

The non-bonding *n* orbital illustrated in Fig. 6.1b has been adapted to the $\mathbf{C}_{2v}$ symmetry of the carbonyl chromophore.

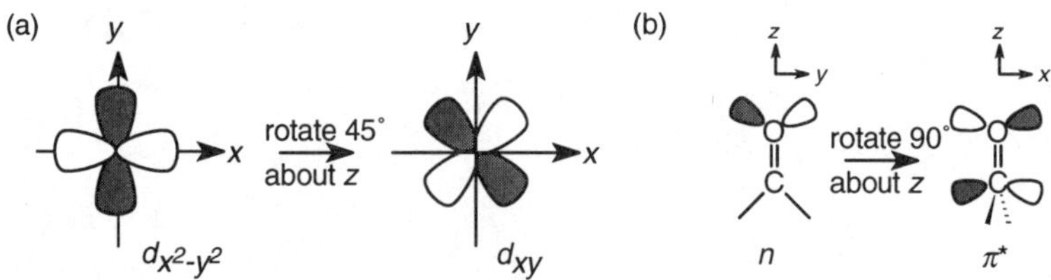

Fig. 6.1 Schematic illustration of (a) a magnetic dipole allowed (mda) *d–d* transition and (b) the $n \to \pi^*$ transition of the carbonyl chromophore. Both transitions involve a rotation of charge. Note rotation of axis system in (b).

The *CD* of the $n \to \pi^*$ transition of carbonyl compounds (Fig. 6.1) has received more attention than that of any other type of transition. The biological importance of many molecules (including steroids) containing the carbonyl chromophore has justified this attention. However, the real reasons are probably more pragmatic. First, their *CD* is comparatively easy to measure since the factor:

$$\frac{\varepsilon_\ell - \varepsilon_r}{\varepsilon_\ell + \varepsilon_r} \tag{6.1}$$

is comparatively large, so the *CD* is an easily detectable percentage of the total absorbance. Second, fairly reliable empirical rules concerning the *CD* sign were deduced at an early stage for the carbonyl $n \to \pi^*$ transition. Another type of mda transition is found among the *d-d* transitions of metal complexes which have also been widely studied. These two systems form the basis of this chapter.

To understand mda transitions let us consider the carbonyl $n \to \pi^*$ transition in a little more detail. During this transition the electrons initially in the n orbital are *rotated* about the z-axis into the π^* orbital. As is illustrated in Fig. 6.1, after rotation there is a net overlap of the positive lobe of the n orbital with that of the π^* orbital, and similarly for the negative lobes. Thus the transition is mda. However, the transition is edf because no linear motion of charge along the symmetry defined axes of the system results in net overlap of positive lobes of the n-orbital with positive lobes of the π^* orbital.

The mdtm is an integral over a product of two factors: the wavefunction describing the final state and the function describing the ground state after the magnetic dipole operator has acted on it. As the effect of the magnetic dipole operator is to rotate the ground state wavefunction, *m* is non-zero if the ground state can be rotated into a function that has net overlap with the excited state.

6.3 $n \to \pi^*$ carbonyl transition and the octant rule

Long before any theoretical understanding of the carbonyl $n \to \pi^*$ transition *CD* had been achieved, it was found empirically that the *CD* of this transition has a sign determined by the factor (Fig. 6.2)

$$-xyz \tag{6.2}$$

where (x, y, z) are the coordinates of a *net perturber* in the framework of the molecule about the carbonyl chromophore. Each atom (or group of atoms) of the molecule induces a contribution to the observed *CD* whose sign is determined by the octant in which the atom lies. However, any atom that has an equivalent 'partner' in a neighbouring octant has its contribution to the *CD* cancelled so it is not a net perturber and may be ignored.

A relationship such as eqn (6.2) where the *CD* sign is related to whether a substituent is in one region of space or another is often referred to as a *sector rule*.

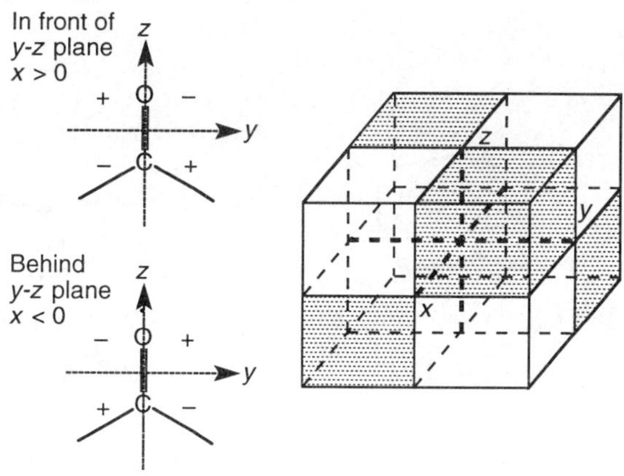

Fig. 6.2 The octant rule.[1]

Hydrocarbon carbonyls and the octant rule

Let us first consider examples (Fig. 6.3a) of carbonyl molecules where, apart from the carbonyl oxygen, all the atoms are Cs or Hs. By convention we ignore the Hs (which by their small polarizability can be considered as very weak perturbers) and consider only the Cs. Building a molecular model or representing the molecule by its projection onto the plane perpendicular to the carbonyl bond (*i.e.* the *x-y* plane) enables the octant sign of any part of a molecule to be determined.

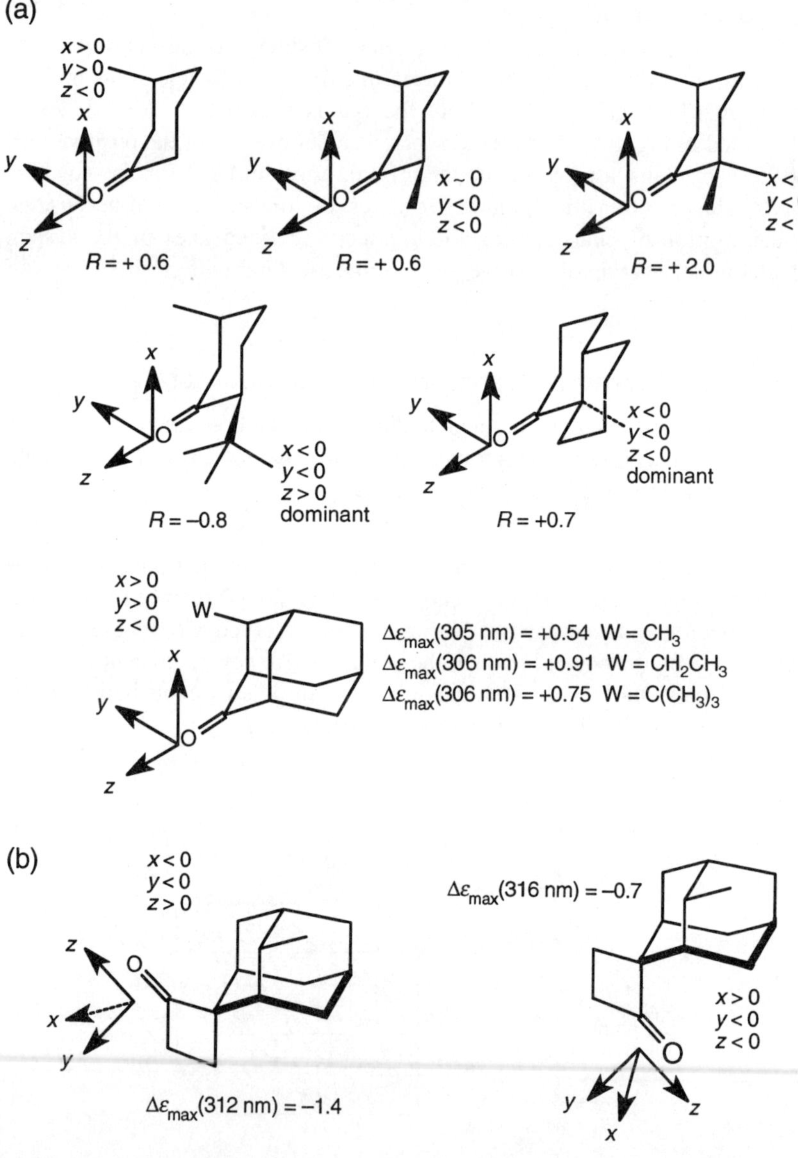

$$\Delta\varepsilon_{max}(305\ \text{nm}) = +0.54 \quad W = CH_3$$
$$\Delta\varepsilon_{max}(306\ \text{nm}) = +0.91 \quad W = CH_2CH_3$$
$$\Delta\varepsilon_{max}(306\ \text{nm}) = +0.75 \quad W = C(CH_3)_3$$

$$\Delta\varepsilon_{max}(316\ \text{nm}) = -0.7$$

$$\Delta\varepsilon_{max}(312\ \text{nm}) = -1.4$$

Fig. 6.3 (a) 'Rear octant' carbonyls. Signs of *x, y,* and *z* indicated for net perturbers, *CD* strengths or $\Delta\varepsilon_{max}$ values are given. (b) Related front and rear octant molecules.[2–4]

Most molecules have all their atoms 'behind' the carbonyl with negative values for z (the so-called 'rear' octants). The molecules of Fig. 6.3a are such rear octant molecules. An example of a 'front' octant molecule is illustrated in Fig. 6.3b.

Heteroatomic carbonyls and the octant rule

A simple series of molecules that illustrate the applicability of the octant rule for heteroatomic carbonyls is provided by the halogenated equatorially substituted adamantanones (Fig. 6.4a). In this case, for $X = CH_3$, Cl, Br, and I as perturbers, the rotatory strength scales with the polarizability of the substituents giving rotatory strengths of +2, +13, +26, and +43 respectively for the 4(S) isomers.[5] This indicates that the *CD* is related to an electronic interaction with the substituent atom rather than to some property related to the motions of the nuclei or steric effects of bulkiness of the substituent groups.

Rotatory strength is proportional to the area under a plot of $\Delta\varepsilon$ versus frequency.

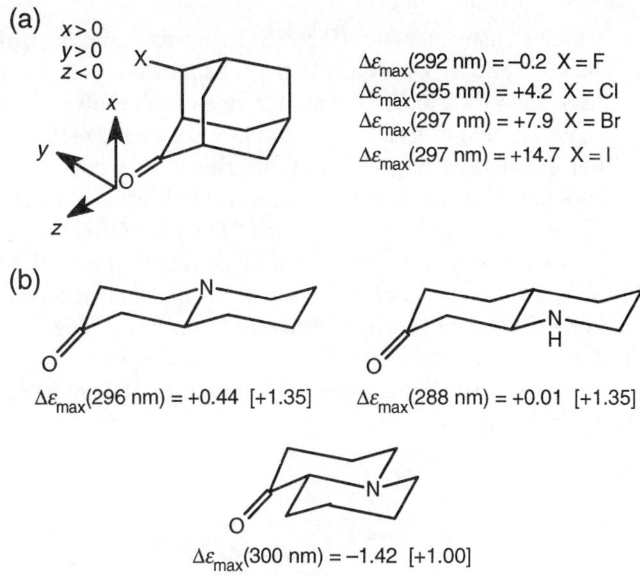

Fig. 6.4 (a) Equatorial adamantanones: (1S,3R)-4(S)(e)-haloadamantanones.[5] (b) Nitrogen containing 'anti-octant' carbonyl compound. The *CD* of the analogous compound with C not N is given in square brackets.[6]

For $X = F$, the rotatory strength for the 4(S) equatorial adamantanone isomer is –0.54, suggesting at first sight a contravention of the octant rule. The same effect is noted when the substituent is deuterium[7] and a number of apparently 'anti-octant' nitrogen-containing compounds have been observed (Fig. 6.4b).[6] The key to this apparent breakdown of the octant rule is provided by the fact that the *CD* strengths scale with substituent polarizability. When there is a deuterium substituent, there is usually a hydrogen in an equivalent position in a neighbouring octant and the polarizability of the C–H bond is larger than that of both the C–D and C–F

D is heavier than H. This causes C–D vibrational energy levels to be lower than those for C–H. This in turn means C–H is more affected by the anharmonicity of the potential energy surface, so its average bond length is longer than that of C–D, so its polarizability is greater.

bonds, thus the H determines the *CD* sign rather than the D or F. Similarly, a nitrogen atom is often over-balanced by a C since the polarizability of C is larger than that of N. Careful inspection of the full molecular structure is thus required before application of the octant rule.

Flexible molecules and the octant rule

In flexible carbonyls, such as substituted cyclohexanones, there is generally a distribution over many different conformations. If one conformation is strongly predominant this may be taken as representative for predictions of the *CD*, although one should recall the possibility that there may be some less abundant conformer that by having a very intense *CD* signal could be important for the net result. A strongly temperature dependent *CD* is often due to slight shifts in equilibria between conformers having widely different *CD*s. Another source of temperature effects may be rearrangement of solvent molecules (see below).

Solvent effects and the octant rule

β-axially substituted adamantanones (Fig. 6.5) have sometimes been observed to give 'anti-octant' effects, especially if the substituent is bulky. Computer simulations have been used to show that in such cases the solvent molecules may be arranged in such a way as to provide a net chiral effect from the solvent.[8,9] For example, a large β-axial substituent can cause a 'solvent deficit' in its octant relative to another octant. Such effects are generally temperature dependent, whereas the normal *CD* of conformationally rigid molecules is essentially temperature independent. Thus the octant contributions of the solvent molecules must be included in any rigorous treatment aiming to account for the observed *CD* spectrum. The importance of solvent effects on electronic spectra has been recognized and today the inclusion of a reaction field calculation in molecular orbital computations of excited states is standard.

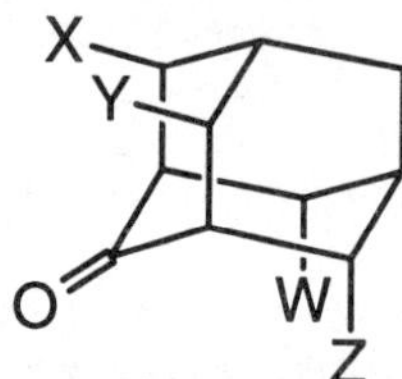

Fig. 6.5 β-axial adamantanones indicating all axial substituent positions.

Perturbers near nodal planes of octants

If the net perturber lies on or very near one of the planes defining the octants, then x or y or z is small, so $-xyz$ will be small. It has been observed then that the sign of the *CD* often does not always correlate with the absolute configuration of the molecule. A number of published papers have endeavoured to show that such a situation requires the central nodal surface through the middle of the C=O bond to be curved. Opposite directions of curvature have been proposed.[2,3] A simpler source for inconsistency between experiment and the octant rule is discussed below when we look at the equations that show why the octant rule works so well most of the time. We shall see that the assumptions required to derive the octant rule break down if $-xyz$ is small.

6.4 $n \to \pi^*$ carbonyl transition and the dynamic coupling model: justification of the octant rule

In order for the $n \to \pi^*$ (or any other mda) transition of an achiral chromophore, **A**, to have *CD* it must acquire electric character in a direction parallel to that of the magnetic dipole transition moment (mdtm). In the previous chapter we saw that eda/mdf transitions, which required induced mdtms, gained their induced moment by coupling with the edtm of another eda transition in a different chromophore. However, magnetic effects are orders of magnitude weaker than electric effects, so the analogous mechanism, where mdtms on different chromophores couple, does not give rise to any detectable *CD*. Instead we need to look for magnetic–electric couplings.

The largest edtm induced into an mda transition results from the coupling of the mda electronic transition with vibrational motions: as **A** vibrates in a normal mode that removes a reflection plane of the carbonyl chromophore, the electron redistribution of the $n \to \pi^*$ transition gains some linear character. The normal absorption intensity of an mda transition is almost wholly due to this mechanism. However, on average, a vibration causes both left-handed and right-handed helices of electron motion, so no *net CD* signal arises due to vibronic coupling.

The largest induced edtm that has *net chirality* comes from the *dynamic coupling* of the mdtm of the $n \to \pi^*$ transition, m^a in **A**, with edtms, μ^c, in chromophores, $\mathbf{C}_i$, that make up the chiral environment of **A**. In order to try to understand the rather complicated equations for the *CD* that are derived in Chapter 7, we express the *CD* strength for any mda transition in terms of an **A** optical factor, $f(\mathbf{A})$, and a **C** optical/geometry factor, $g(\mathbf{C})$ (which includes the energy terms):

$$R = f(\mathbf{A})g(\mathbf{C}) \tag{6.3}$$

The first non-zero term of the *CD* strength for the carbonyl $n \to \pi^*$ transition has an **A** optical factor

$$f(\mathbf{A}) = \mathrm{Im}\left[Q^a_{xy} m^a_z\right] \tag{6.4}$$

and a **C** optical/geometry factor:

$$g(\mathbf{C}) = \frac{-6\varepsilon_c}{\left(\varepsilon_c^2 - \varepsilon_a^2\right)R_{AC}^4} \, \mu_z^c \left[\mu_x^c \hat{R}_{ACy} + \mu_y^c \hat{R}_{ACx} - \sum_k^{x,y,z} 5\mu_k^c \hat{R}_{ACx} \hat{R}_{ACy} \hat{R}_{ACk}\right] \tag{6.5}$$

where ε_c is the transition energy for **C** transition moments μ^c, ε_a is the transition energy for the $n \to \pi^*$ transition which has mdtm m^a and electric quadrupole transition moment Q^a. Q^a_{xy} is the x-y component of Q^a. R_{AC} is the vector from the origin of chromophore **A** to the origin of chromophore **C**

If there is more than one **C** chromophore then we introduce a sum over $\mathbf{C}_i$
$$R = \Sigma_i f(\mathbf{A})g(\mathbf{C}_i)$$

The quadrupole transition moment Q^a is a second rank tensor, it thus has two dimensions.

as in the previous chapter, and $\hat{R}_{AC}$ is the unit vector along that line (Fig. 6.6).

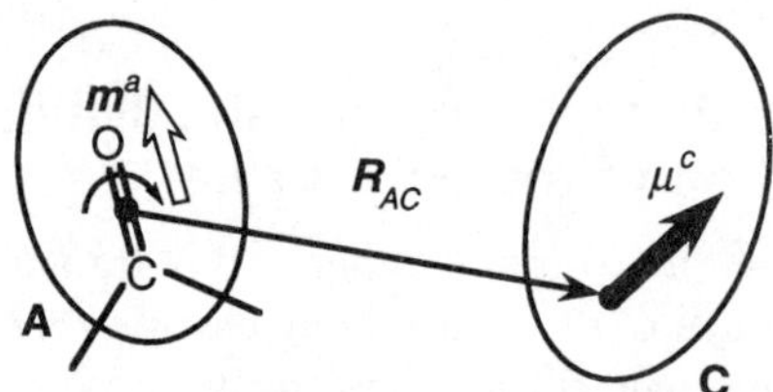

Fig. 6.6 Geometry of a carbonyl **A**/**C** system (*cf.* Fig. 5.6).

The definition of the chromophores of a carbonyl molecule

The achiral chromophore **A** is the carbonyl chromophore composed of the C=O bond and the two carbon atoms to which it is joined. The rest of the molecule is divided into chromophores, $\mathbf{C}_i$, which are subunits of the molecule that contain electronic transitions (represented by transition moments or polarizabilities) and do not exchange electrons with the rest of the molecule. For a molecule where (apart from the carbonyl group) all the valence electrons are in sigma bonds, the valence electron density of the molecule is localized in two-centre two-electron bonds between pairs of atoms. For such molecules taking the $\mathbf{C}_i$ to be bonds is a good approximation. Alternatively, the $\mathbf{C}_i$ may be defined to be atoms such as carbons and nitrogens. Although the bond approximation is the better one, in most cases the difference in xyz for the middle of a bond and one of the terminal atoms of the bond is not great.

If there are π-bonds or non-bonding electrons in the molecule, then the unit where the electrons are localized is often somewhat larger than a single bond between two atoms, *e.g.* if the molecule has a $-$COOH, $-$NH$_2$, or $-$C$_6$H$_5$ group, these will need to be treated as units. The correct identification of spectroscopic chromophores can usually be achieved by dividing the molecule into chemical functional groups.

The octant rule

To derive the octant rule [eqn (6.2)] from eqns (6.3)–(6.5), we take the $\mathbf{C}_i$ to be the atoms of the molecule, then replace *e.g.* R_{ACx} by x for each atom. Then ignore all terms of eqn (6.5) except the final one with $k = z$ as this is the isotropic polarizability component. The octant rule then follows. The assumptions made in this derivation are discussed in §7.9.[10]

$$R \approx \frac{30\varepsilon_c}{\left(\varepsilon_c^2 - \varepsilon_a^2\right)R_{AC}^4} f(\mathbf{A})\mu_z^c\mu_z^c\left[\hat{R}_{ACx}\hat{R}_{ACy}\hat{R}_{ACz}\right]$$

$$\approx \frac{30\varepsilon_c}{\left(\varepsilon_c^2 - \varepsilon_a^2\right)R_{AC}^4} f(\mathbf{A})\mu_z^c\mu_z^c\left[xyz\right] \tag{6.6}$$

$$\approx \frac{30\alpha(\varepsilon_a)}{R_{AC}^4} f(\mathbf{A})\left[xyz\right]$$

where the dynamic polarizability at energy ε_a is

$$\alpha_c(\varepsilon_a) = \frac{1}{3}\sum_c \frac{\varepsilon_c \boldsymbol{\mu}^c \cdot \boldsymbol{\mu}^c}{\varepsilon_c^2 - \varepsilon_a^2} \qquad (6.7)$$

As $\varepsilon_c > \varepsilon_a$, the octant rule requires $f(\mathbf{A})$ to be negative for carbonyls.

In practical applications of the octant rule, the groups for which the product xyz is determined are sometimes atoms, sometimes bonds, and sometimes functional groups (usually equivalent to spectroscopic chromophores). The main assumptions made in going from eqn (6.5) to eqn (6.6) are that the diagonal components of the dynamic polarizability of each $\mathbf{C}_i$ are much larger than the off-diagonal components (the uv component of the polarizability describes how the electrons move in the v direction if they are pushed in the u direction) and xyz is not close to zero. If xyz is close to zero, then one of the first two terms in the square brackets of eqn (6.5) may be larger than the second. This is one of the reasons why the octant rule breaks down when the net perturber lies close to one of the planes defining the octants (see above). We have further simplified eqn (6.5) by using an average of the diagonal polarizability terms by assuming that $\mu^2 \approx 3\mu_z^2$.

Two features of the octant rule that are omitted from eqn (6.2) follow from eqn (6.6). The first is the already noted dependence on net perturber polarizability. The second is the R_{AC}^{-4} distance dependence of the CD expression. This enables us to assess the relative contributions of two or more net perturbers should they be present in a carbonyl molecule.

Thus, to apply the octant rule in the revised form indicated by eqn (6.6) first identify the chromophores, $\mathbf{C}_i$, into which the non-carbonyl part of the molecule is to be divided (usually sigma bonds and functional groups). Second, determine for each $\mathbf{C}_i$

- its isotropic polarizability[8]
- its distance from the carbonyl chromophore origin (which we take to be the centre of the C=O bond)
- the unit vector from the carbonyl origin to the $\mathbf{C}_i$ origin.

Eqn (6.6) may then be evaluated for each $\mathbf{C}_i$ in terms of the parameter $f(\mathbf{A})$ which we assume is negative and constant for all ketone carbonyls.

Electronic causes for failures of the octant rule

The octant rule may break down when the carbonyl group cannot be regarded as an **A**-chromophore that is spectroscopically isolated from the rest of the molecule. Some steroids, such as progesterone and testosterone (Fig. 6.7), are cases in point, since the carbonyl is conjugated with the carbon framework of the molecule.

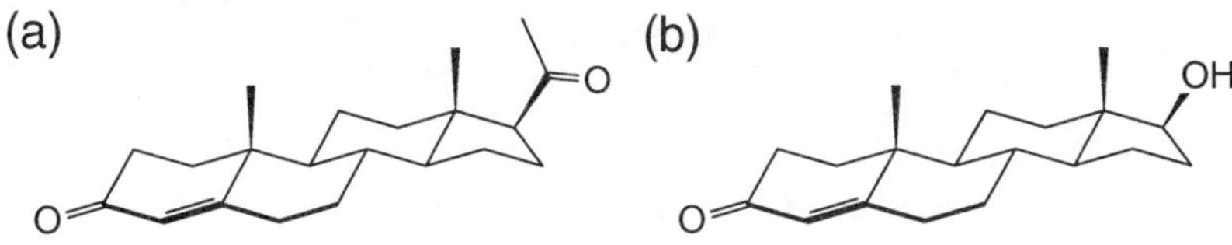

Fig. 6.7 (a) Progesterone and (b) testosterone.

Substituents linked to a carbonyl chromophore by a 'planar zig-zag' chain often induce a larger *CD* signal than expected from eqn (6.6) on the basis of their polarizability and distance from the carbonyl. They also cause the transition energy to be red-shifted relative to the same substituent in other positions. The reason probably lies in the overlap of the n orbital with orbitals of the bonds of the zig-zag. The interaction between the n orbital and the σ antibonding orbitals of the zig-zag is illustrated in Fig. 6.8.[11] The zig-zag may be thought of as providing an enhanced communication of the substituent's polarizability-induced helical twisting of the $n \rightarrow \pi^*$ transition. Similarly, σ orbitals on Z2, Z4 (Fig. 6.8) interact with the π^* orbital of the carbonyl.

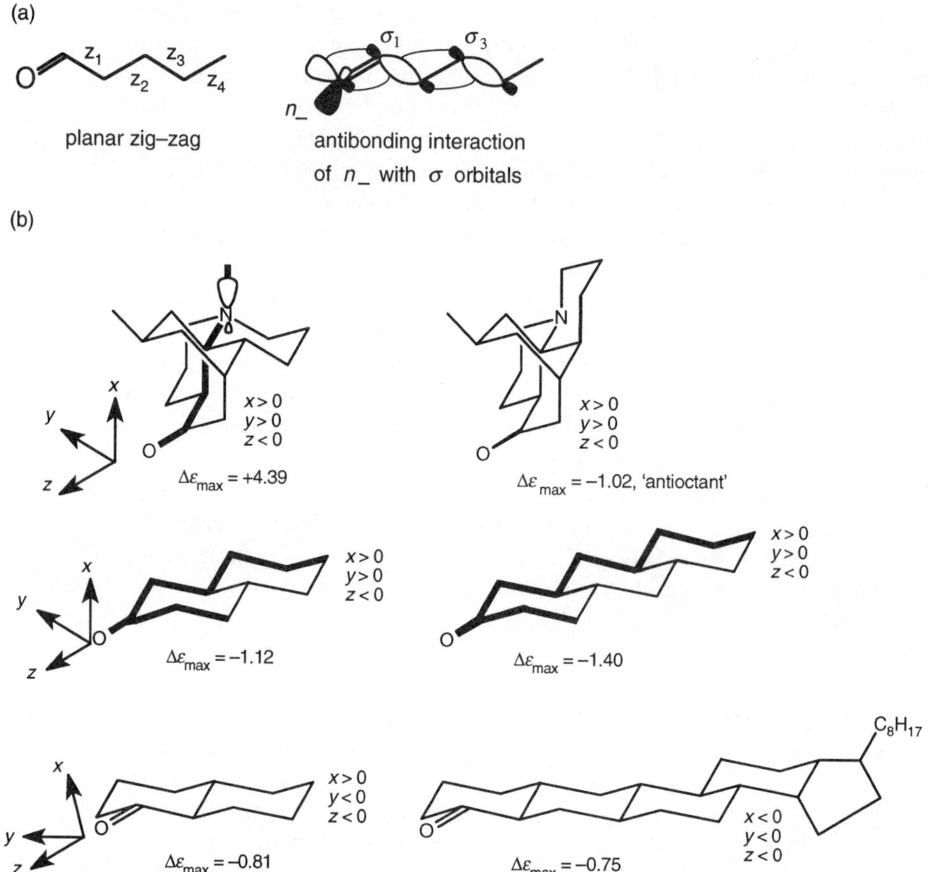

Fig. 6.8 (a) Antibonding interactions of n_- and σ orbitals in planar zig-zag carbonyls.[11] (b) Planar zig-zag molecules (zig-zags shown by thick lines) where an enhanced red-shifted *CD* is observed. Related non-zig-zag cases are also illustrated.[6,12]

6.5 *d-d* transitions of transition metal complexes: symmetry dependence of the *CD* of mda transitions

Another class of mda/edf transitions that have been extensively studied are the *d-d* transitions of transition metal complexes (Fig. 6.1a). These transitions usually occur at lower energies than the charge transfer and in-ligand transitions that we examined in Chapter 5. They are therefore clearly identifiable in the spectrum even though the extinction coefficients of edf transitions are generally orders of magnitude smaller than those of eda transitions. In contrast to the situation for the metal complex eda transitions studied in Chapter 5 where we could describe the *CD* for any molecule in terms of degenerate and non-degenerate coupled-oscillator equations, with the mda/edf *d-d* transitions the form of the *CD* equation depends both on the symmetry of **A** and on the polarization of the intrinsic magnetic dipole transition moment of the mda transition.

The equations for the *CD* of the $n \rightarrow \pi^*$ carbonyl transition are appropriate for the z-polarized mda *d-d* transitions (the mdtm lies along z) of metal complexes with $\mathbf{C_{2v}}$ symmetry **A** chromophores (where **A** is the metal and directly ligating atoms) so they follow the octant rule. For x- and y-polarized mda *d-d* transitions of such $\mathbf{C_{2v}}$ systems the *CD* is given by:

$$R(x) = \frac{6\alpha_c(\varepsilon_a)}{R_{AC}^3} \operatorname{Im}\left[\mu_y^a m_x^a\right][xy] \tag{6.8}$$

$$R(y) = \frac{6\alpha_c(\varepsilon_a)}{R_{AC}^3} \operatorname{Im}\left[\mu_x^a m_y^a\right][xy] \tag{6.9}$$

The geometry dependence of eqns (6.8) and (6.9) is a quadrant, xy, instead of an octant rule and the dependence on the **A**–**C** distance is inverse third rather than inverse fourth power (Fig. 6.9). Thus the *CD* of an x- or y-polarized *d-d* transition is expected to be larger than that for a z-polarized transition.

Under $\mathbf{C_{2v}}$ symmetry, transitions with x- and y-polarized mdtms actually have edtms of, respectively, y and x polarization. However, the eda coupled-oscillator *CD* (Chapter 5) is not significant here as the edtms are very small—they are zero under the higher symmetry of the octahedral template.

Molecule	A symmetry	R_{AC}^{-n}	Isotropic geometry factor
$\mathbf{D_3}$	$\mathbf{O_h}$	$n = 6$	0
$\mathbf{C_3}$	$\mathbf{D_{3d}}$	E: $n = 4$ A$_2$: $n = 6$	$x^3 - 3xy^2$ $(x^3 - 3xy^2)(9z^2 - 1)$
$\mathbf{C_2}$	$\mathbf{C_{3v}}$	E: $n = 3$ A$_2$: $n = 5$	$x^3 - 3xy^2$ $(x^3 - 3xy^2)(9z^2 - 1)$
$\mathbf{C_2}$	$\mathbf{C_{4v}}$	E: $n = 3$ A$_2$: $n = 6$	xy $(x^2 - y^2)xyz$
$\mathbf{D_4}$	$\mathbf{D_{4h}}$	E: $n = 4$ A$_2$: $n = 6$	$0, \rightarrow n = 6$ $(x^2 - y^2)xyz$
$\mathbf{C_2}$	$\mathbf{C_{2v}}$	B: $n = 3$ A$_2$: $n = 4$	xy xyz

Fig. 6.9 Distance dependence and geometry factors for the isotropic polarizability term in equations describing the *CD* induced into *d-d* chromophores of different symmetry. Circles and squares denote different ligating atoms.

We are considering mda/edf *d-d* transitions here. Some *d-d* transitions are mdf/edf. The latter are derived from T_{2g} symmetry transitions in octahedral metal complexes, whereas the mda transitions come from T_{1g} transitions.

Unfortunately, it is seldom possible to use this *CD* strength magnitude difference to help assign *d-d* transitions for $\mathbf{C}_{2v}$ systems. The reason for this lies in the nature of the transition and the high template symmetry of the molecules. The fact that we can label the *d-d* transitions as such indicates that the transitions are between fairly pure *d* orbitals. Thus, as in the higher symmetry octahedral complexes such as $[Co(NH_3)_6]^{3+}$, x-, y-, and z-polarized transitions occur very close together in energy (they are degenerate for octahedral complexes) and so a significant amount of cancellation will occur between overlapping *CD* bands of opposite signs.

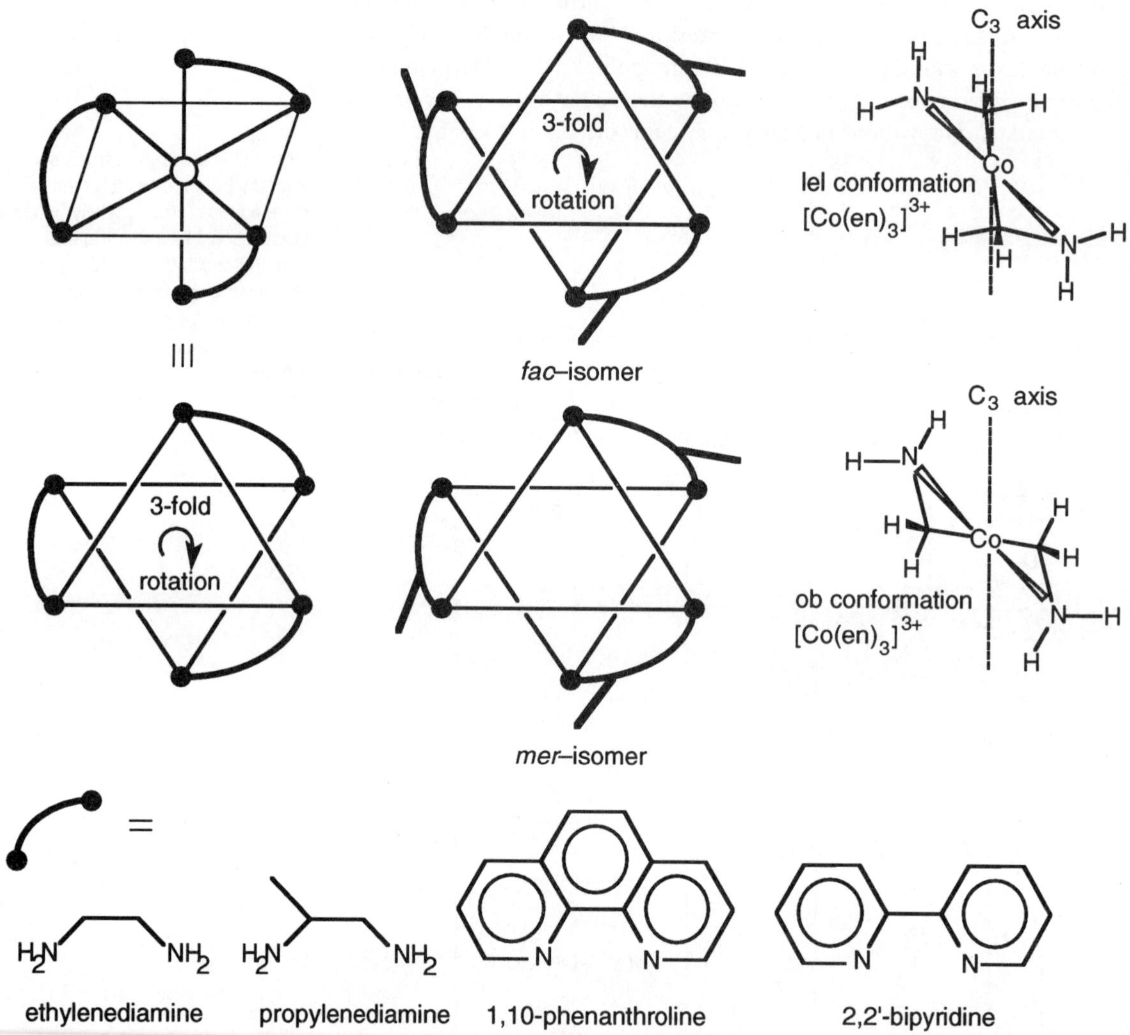

Fig. 6.10 Δ-enantiomer tris-chelate transition metal complexes illustrating the geometry of the *fac* and *mer* isomers adopted by [Co(propylenediamine)₃]³⁺ and the *lel* and *ob* conformers of [Co(ethylenediamine)₃]³⁺.

When we wish to consider higher symmetry molecules, such as the tris-chelate complexes of Fig. 6.10 it proves necessary to go to higher order **A** moment products and much more complicated geometry dependences. Some possibilities are summarized in Fig. 6.9. Here we shall focus on tris-chelate complexes, for which the *d-d* chromophore **A** has approximately $\mathbf{D}_{3d}$ symmetry. We therefore expect two transitions close together in energy, a *z*-polarized A_2 symmetry transition and a degenerate *x,y*-polarized E symmetry transition.

The isotropic polarizability terms of the *CD* equations for *d-d* transitions of tris-chelates are:[13]

$$R(\mathrm{E}) = \frac{15\alpha(\varepsilon_a)}{4R_{AC}^4}\,\mathrm{Im}\left[m_x^a Q_{x^2}^a - m_x^a Q_{y^2}^a - 2m_y^a Q_{xy}^a \right]\left[x\left(x^2 - 3y^2\right)\right] \quad (6.10)$$

$$R(\mathrm{A}_2) = \frac{35\alpha(\varepsilon_a)}{4R_{AC}^6}\,\mathrm{Im}\, m_z^a\left[H_{x^3z}^a - 3H_{xy^2z}^a \right]\left[x\left(x^2 - 3y^2\right)\left(9z^2 - 1\right)\right] \quad (6.11)$$

where $Q_{x^2}^a$ etc. and $H_{x^3z}^a$ etc. are, respectively, quadrupole and hexadecapole transition moments. The distance dependence of these equations would lead us to expect the E band *CD* to be larger than the A_2 band. The 'dominant E band rule' was in fact first postulated as an empirical rule based on the study of a significant number of systems and is illustrated by the spectra in Figs 1.5 and 6.11.

There has been some controversy about the dominant E band rule: when the *CD* spectrum of crystalline [Co(ethylenediamine)$_3$]$^{3+}$ was measured it was found that the crystal *CD* for the A_2 component of the T_{1g} *d-d* band had almost the same magnitude as that of the E band and both were an order of magnitude larger than those observed in the solution spectrum—so the solution spectrum appeared to be a result from the cancellation between two large A_2 and E bands. This raises the question of why the E band in solution spectra is always the larger if the observed solution *CD* is only a small percentage of the total crystal signal.

A possible resolution of this problem is to conclude that the conformation of the ethylenediamine rings in solution is different from that in the crystal. [Co(propylenediamine)$_3$]$^{3+}$ (Fig. 6.10), however, is forced to adopt the crystal conformation (*lel*) in solution due to its extra methylene group, and although the A_2 component of its *CD* (Fig. 6.11) is larger than for [Co(ethylenediamine)$_3$]$^{3+}$ (Fig. 1.5), the E component is clearly dominant.

As always, to resolve the apparent contradiction between a calculation and experiment we must examine the assumptions underlying the calculation. The first assumption made in deriving eqns (6.10) and (6.11)[13] was that the *d-d* chromophore is achiral. As soon as one considers the crystal coordinates one realizes that the *d-d* chromophore is not achiral—there is always an element of twist. The assumption of achiral **A** symmetry may, however, be valid in solution due to atomic motions of the molecule leading to an averaging of the twist. Thus, we might expect an intrinsic *CD* contribution for the lower symmetry crystal situation that is absent from the solution spectrum.

Alternatively this problem may be explained in terms of a difference between *additivity* and *coalescence* effects. Let us consider the hypothetical experiment of gradually changing the geometry of **A** so that the symmetry of the complex increases. It is intuitively clear that the magnitude of *CD* bands

We saw the effect of additivity when we considered the coupled-oscillator *CD* for when the interaction energy, *V*, was zero (§5.4). That the octant rule does not contribute to *tris*-chelate *CD* is an example of coalescence.

will continuously decrease with increasing symmetry until, when the system becomes achiral, the *CD* vanishes. Although symmetry changes may be abrupt, the geometry changes involved can be thought of as continuous so there is a tendency to imagine bands belonging to transitions of different polarizations merging to form one band as transitions become degenerate. However, two distinct factors are operative, both of which may contribute to the large difference between crystal and solution *CD*.

- *Additivity* decribes the cancellation of *CD* intensity observed when transitions occurring at different energies in low symmetry systems are degenerate in higher symmetry systems. The individual *CD* strengths, which derive from coupling with other transitions (in different chromophores) are the same for the low and high symmetry systems.

- *Coalescence* describes what occurs when increased symmetry causes *CD* mechanisms that contribute to the *CD* of lower symmetry systems to vanish, so the net *CD* decreases. This effect corresponds to the situation of an isolated chiral chromophore in which the *CD* of all transitions should sum to zero. To a reasonable approximation, a local sum rule also holds for the *CD* of *e.g.* the two transitions of a $\mathbf{D}_{3d}$ chromophore deriving from the T_{2g} band.

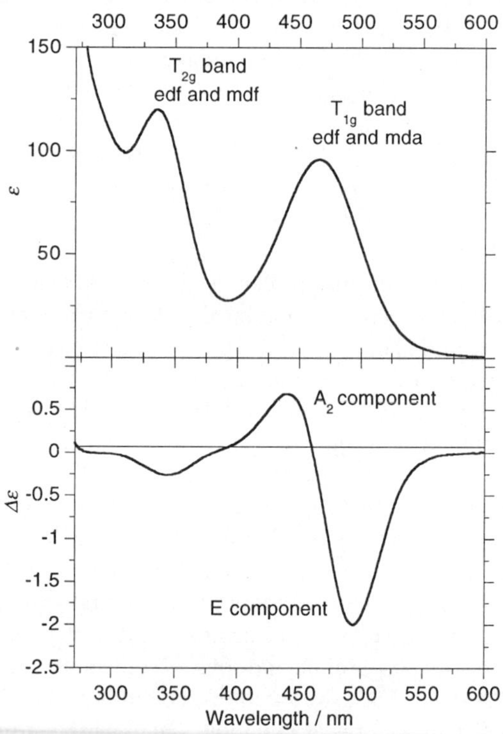

Fig. 6.11 *d-d CD* spectrum of Δ-[Co(R-propylenediamine)$_3$]$^{3+}$.

In deriving eqns (6.10) and (6.11) for tris-chelate *CD* it is assumed that **A** has $\mathbf{D}_{3d}$ symmetry. This implicitly performs an additivity cancellation for the A_2 and E band *CD* intensity. In particular it cancels any *CD* due to the intrinsic chirality of the *d-d* chromophore that we have ignored. In solution this does not matter because the small energy splitting between the A_2 and E bands also performs this additivity cancellation. However, in measuring the single crystal *CD*, only the E band is measured so no automatic additivity cancellation occurs. In both instances, however, the *net CD* is dominated by the mechanisms of eqns (6.10) and (6.11).

6.6 Magnetic circular dichroism

The final section of this chapter is devoted to magnetic circular dichroism, *MCD*. An *MCD* spectrum is measured by first placing the sample in a magnetic field that is aligned parallel with (by convention the direction of the magnetic field is north to south) the direction of propagation of light and then measuring the *CD* spectrum (Fig. 6.12).

MCD signal magnitudes are dependent on the magnitude of the magnetic field, but the large magnetic fields used for NMR and mass spectroscopy are not used for *MCD*.

In practice, most *MCD* signals are much smaller than intrinsic *CD* signals, so *MCD* is only ever measured for achiral molecules. *MCD* is most commonly used to see how many transitions are under one absorption envelope, since transitions of different polarizations often have differently signed *MCD* signals. *MCD* spectroscopy has never become a general chemical tool, probably because the theory looks very complicated.

MCD is becoming more and more widely used in the field of bioinorganic chemistry.[14]

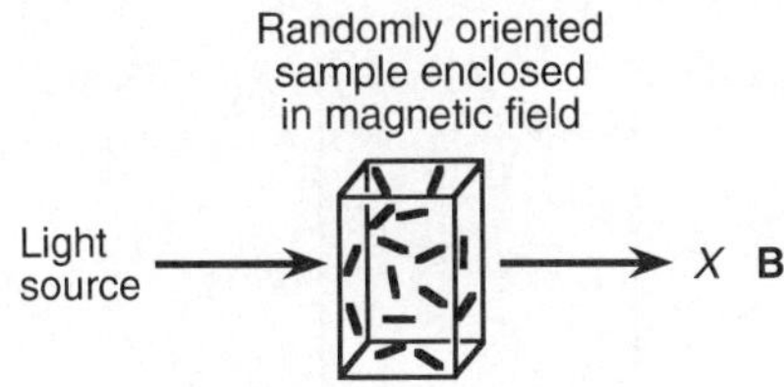

Fig. 6.12 Configuration of an *MCD* experiment.

Our treatment of *MCD* which is derived in Chapter 7 is much simpler than that usually found in the literature[14,15] since the latter approach traditionally explicitly includes the band shapes into derivations and calculations. The advantage of including band shapes from the beginning is that the final equations relate directly to the shapes of observed spectra, however, to a considerable extent this is at the expense of a simple physical understanding of the origin of the measured *CD*.

Our approach to deriving *MCD* equations is analogous to the coupled-oscillator analysis of Chapter 5. The main difference between the two situations is that in the coupled-oscillator model the coupling is between an eda transition and the magnetic dipoles generated by other transitions on neighbouring chromophores whereas with *MCD* the *CD* arises from the coupling between an eda transition and the externally applied magnetic field *via* another eda transition on the same chromophore.

Now consider a randomly oriented collection of molecules in a magnetic field of magnitude B that is directed along X, the direction of propagation of the incident radiation. The *MCD* of a non-degenerate transition with edtm μ^{01} and transition energy ε_1 is

$$MCD = \frac{4k\text{BIm}\left(\mu^{01} \times \mu^{02} \cdot m\right)}{3\left(\varepsilon_2 - \varepsilon_1\right)} \tag{6.12}$$

where k is a constant (see Chapter 7), Im denotes 'imaginary part of', μ^{02} is a neighbouring transition occurring at energy ε_2, and m is the magnetic dipole transition moment *between the excited states* from state $|1)$ to state $|2)$. For a non-zero *MCD* signal μ^{01} and μ^{02} must therefore not be parallel. If there is more than one state close in energy to the one for which we are measuring the *MCD*, then a sum over these states must be included in eqn (6.12).

The *MCD* for the neighbouring transition (μ^{02}) follows from eqn (6.12) upon doing three things, each of which changes the sign of the expression so the final result is an *MCD* signal of opposite sign from that of eqn (6.12):

- exchange μ^{01} and μ^{02}
- exchange ε_1 and ε_2
- reverse the direction of m.

The last sign reversal follows from exchanging the wavefunctions in the expression for the magnetic transition moment (see §7.10).

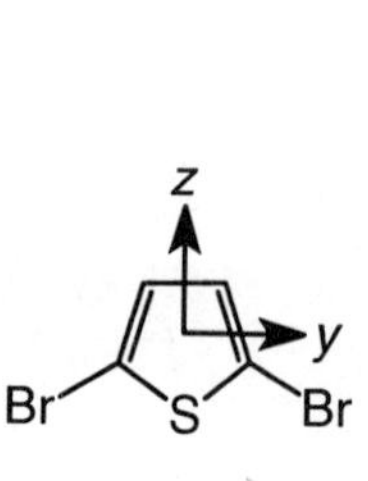

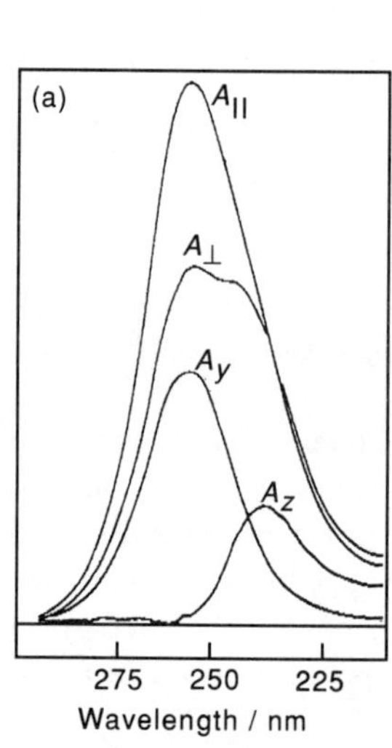

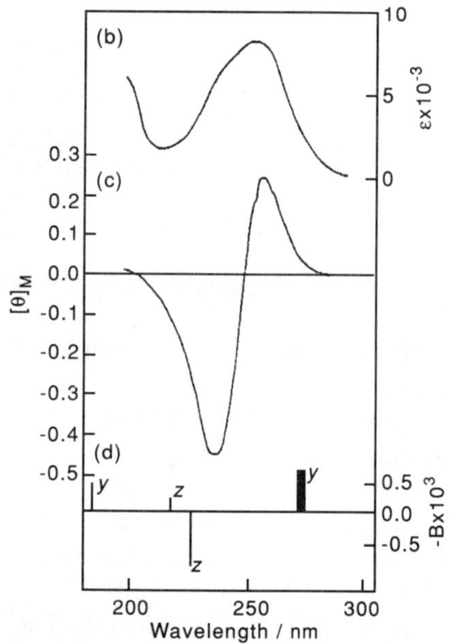

Fig. 6.13 *MCD* spectrum of dibromothiophene showing opposite signed signals for neighbouring transitions of perpendicular polarization.[16]

To sum up, the *MCD* signal for a non-degenerate transition arises from coupling with another transition that is fairly close in energy and whose transition polarization is different. The *MCD* signal of the second transition (assuming coupling to all other transitions can be ignored) will be equal in magnitude, but opposite in sign from that of the first. Fig. 6.13 gives the *MCD* spectrum for dibromothiophene, together with the corresponding resolved stretched film *LD* spectra showing the presence of two orthogonal, polarized absorption components. In the unsubstituted thiophene (as well as in its seleno or telluro analogues) the two transitions are near degenerate. As an effect of the small energy difference in the denominator of eqn (6.12), therefore, the *MCD* signals are larger than for dibromothiophene (Fig. 6.14) where the transitions are further apart from each other.

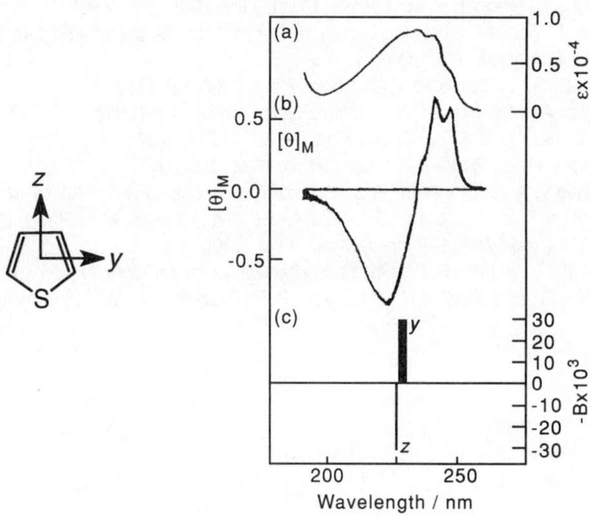

Fig. 6.14 *MCD* spectrum of accidentally degenerate transitions in thiophene.[16]

If the transition of interest, with edtm $\boldsymbol{\mu}^{01}$ and energy ε, goes from a non-degenerate ground state to a degenerate excited state (of symmetry E or T), then there will be another transition, with edtm $\boldsymbol{\mu}^{02}$ occurring at the same energy. There will then be two *MCD* bands with intensity

$$MCD(\pm) = \pm k\boldsymbol{\mu}^{01} \times \boldsymbol{\mu}^{02} \cdot \hat{\boldsymbol{B}} \qquad (6.13)$$

at energies

$$\varepsilon^{\pm} = \varepsilon \mp \mathrm{Im}(\mathbf{B} \cdot \boldsymbol{m}) \qquad (6.14)$$

Note: $\hat{\mathbf{B}} = (1,0,0)$, and the choice of sense for $\boldsymbol{\mu}^{01}$ and $\boldsymbol{\mu}^{02}$ is arbitrary, but they define the sense of $\boldsymbol{m}$.

In eqn (6.13) the *MCD magnitude* does not formally depend on the magnitude of the magnetic field, but eqn (6.14) which gives the energy splitting between the transitions does. This is similar to the situation with the degenerate coupled-oscillator model, and an analogous exciton-like spectrum is therefore expected for such a situation.

Although the underlying mechanisms of all *MCD* spectra are in principle the same, the equations for different cases have by convention been given specific labels: the *MCD* from non-degenerate couplings is known as the 'B term' while that for couplings of degenerate excited states is the 'A term'. The situation with a degenerate ground state, where the magnetic field lifts the degeneracy and there is an asymmetric thermal population distribution, gives rise to the 'C term'.[14,15]

References

(1) Moffitt, W.; Woodward, R. B.; Moscowitz, A.; Klyne, W.; Djerassi, C. *J. Amer. Chem. Soc.* **1961**, *83*, 4013

(2) Coulombeau, C.; Rassat, A. *Bull. Soc. Chim. de France*, **1971**, 2516

(3) Lightner, D. A.; Bouman, T. D.; Wijekoon, W. M. D.; Hansen, A. *J. Amer. Chem. Soc.* **1985**, *89*, 5805

(4) Lightner, D. A.; Chang, T.C. *J. Amer. Chem. Soc* **1974**, *96*, 3015

(5) Snatzke, G.; Eckhardt, G. *Tetrahedron*, **1968**, *24*, 4543

(6) Hudec, J. *J. Chem. Soc. Chem. Commun.* **1970**, 829

(7) Lightner, D. A.; Gawronski, J.K.; Bouman, T.D. *J. Amer. Chem. Soc.* **1980**, *102*, 1983

(8) Rodger, A.; Rodger, P. M. *J. Amer. Chem. Soc.* **1988**, *110*, 2361

(9) Fidler, J.; Rodger, P. M.; Rodger, A. *J. Chem. Soc. Perkin II* **1993**, 235; *J. Amer. Chem. Soc.* **1994**, *116*, 7266

(10) Höhn, E. G.; Weigang, O. E. *J. Chem. Phys.* **1968**, *48*, 1127

(11) Rodger, A.; Moloney, M. G. *J. Chem. Soc. Perkin II* **1991**, 919

(12) Kirk, D. N.; Klyne, W. *J. Chem. Soc. Perkin I* **1974**, 1076

(13) Schipper, P. E.; Rodger, A. *Chem. Phys.* **1986**, *109*, 173

(14) McCaffery, A. J.; Stephens, P. J.; Schatz, P. N. *Inorg. Chem.* **1967**, *6*, 1614; Stephens, P. J. *Adv. Chem. Phys.* **1976**, *35*, 197; Solomon, E. I.; Pavel, E. G.; Loeb, K. E.; Campochiaro, C. *Coord. Chem. Rev.* **1995**, *144*, 369

(15) Michl, J.; Thulstrup, E. W. *Spectroscopy with polarized light*. New York: VCH, **1986**

(16) Nordén, B.; Håkansson, R.; Pedersen, P. B.; Thulstrup, E. W. *Chem. Phys.* **1978**, *33*, 355

7 Circular dichroism formalism

7.1 Introduction

This chapter is designed to enable readers to understand the basis of the equations used in previous chapters and to enable the *CD* theory literature to be read. Much of the interpretation of the final equations has already been covered in the previous chapters. This chapter therefore has very few pictures, comparatively little discussion, but many equations.

The level of mathematical sophistication required for this chapter is comparatively low, but the necessity of labelling transition moments and often transition energies with the states they connect and the chromophore to which they belong makes many of the equations appear complicated. The mathematics required includes: a basic acceptance of quantum theory (*i.e.* molecules have wavefunctions that describe their electron distribution), the use of *bra-ket* notation, vector products, trigonometry, complex numbers, and simple determinants. Brief textual and marginal notes endeavour to provide sufficient detail on these topics to remind readers who have previously covered them but have forgotten the details. For readers who find these notes too brief, most physical chemistry textbooks or mathematics-for-chemists books will give more details.

The chapter is structured so the next three sections cover general aspects of the interaction of polarized radiation and matter. The emphasis of the final sections of the chapter is on determining wavefunctions and hence *CD* expressions for a transition in chromophore **A** when perturbed by the rest of the molecule or by an external magnetic field. In §7.5 and §7.6 the focus is on transitions that to a first approximation only have linear charge displacements (*i.e.* eda and mdf transitions). Circular charge displacements are covered in §7.8. Each type of transition gains the helical character required for a *CD* signal by perturbation from another part of the system. In the final section we shall see the effect of an external static magnetic field on eda transitions of achiral molecules when we derive the *MCD* equations of §6.6. Each section needs to be cross referenced with its 'application' section in previous chapters.

7.2 Polarized light and spectropolarimeters

Polarized light

As discussed in §1.2, linearly polarized light beams have all the photons with their electric fields oscillating in the same plane (Fig. 1.3). When two parallel linearly polarized beams are combined this will only result in a new linearly polarized light beam if the phases (and wavelengths) of both beams are the same—*i.e.* if both beams have zero amplitude at the same points in space. If we take X to be the direction of propagation of the light, then the unit vectors describing the polarization (by convention the direction of the electric field component of the light) of two in-phase (*i.e.* both light beams have zero amplitude at the same points in time and space) perpendicularly polarized light beams may be written

$$\hat{\mathbf{e}}_1 = \hat{\mathbf{e}}_Y = (0,1,0) \tag{7.1}$$

and

$$\hat{\mathbf{e}}_2 = \hat{\mathbf{e}}_Z = (0,0,1) \tag{7.2}$$

Any linear combination of these two in-phase beams will be another linearly polarized light beam. If, however, the two beams are out of phase, the result of combining them will be elliptically polarized light—*i.e.* the electric field of the light will trace out an ellipse.

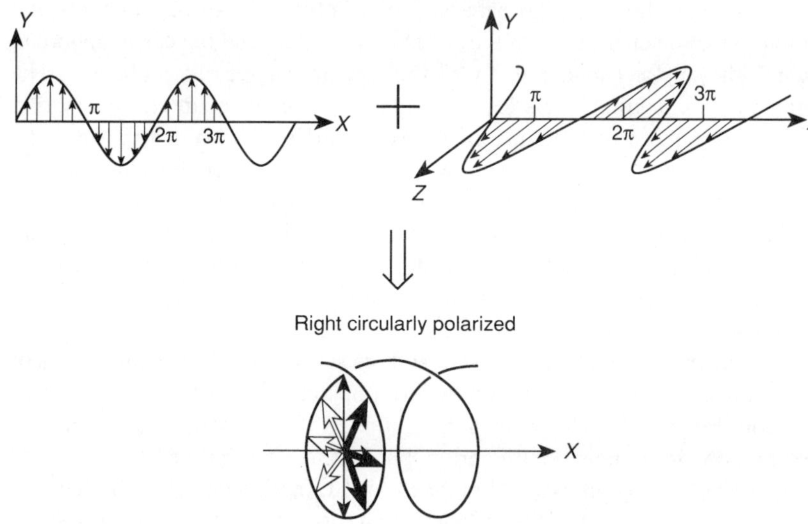

Fig. 7.1 Production of circularly polarized light from two linearly polarized light beams. Arrows indicate electric field polarizations.

In the special case where two equal magnitude linearly polarized beams of orthogonal polarizations are combined with a phase difference of $\pi/2$ (quarter of a wavelength) we get circularly polarized light. The electric field polarization vector for right circularly polarized light may be written:[1,2]

$$\hat{\mathbf{e}}_r = \frac{1}{\sqrt{2}}\left\{(0,1,-i)\exp\left(\frac{2\pi iX}{\lambda} - i\omega t\right)\right\} \tag{7.3}$$

In using this definition of right (and hence left) circularly polarized light we are following the optics convention.[1,2] If one plots a helix about a time axis, t, right circularly polarized light forms a left-handed helix in time.

where i is the square root of -1. The oscillation of the electric field vector in time is described by the factor $\exp(-i\omega t)$ where ω is the angular frequency $(2\pi\nu)$ and t is time: the magnitude of the electric field vector remains constant in time, but its direction rotates about X with frequency ω. By taking the real part of eqn (7.3) we see that for an observer sitting at a fixed point, say $X = 0$, the electric field vector of right circularly polarized light rotates in time about X in a clockwise manner. At a fixed time, say $t = 0$, it forms a right-handed helix in space. In what follows we use the polarization vectors:

$$\hat{\mathbf{e}}_\ell = \frac{1}{\sqrt{2}}(0,1,i)$$

$$\hat{\mathbf{e}}_r = \frac{1}{\sqrt{2}}(0,1,-i)$$

(7.4)

Circular dichroism spectropolarimeter

As *CD* is the difference in absorption of left and right circularly polarized light, the key feature of a *CD* spectropolarimeter is a means of producing both polarizations of light with exactly equal intensities. The light source in most instruments is a xenon arc lamp for UV and visible *CD* measurements.

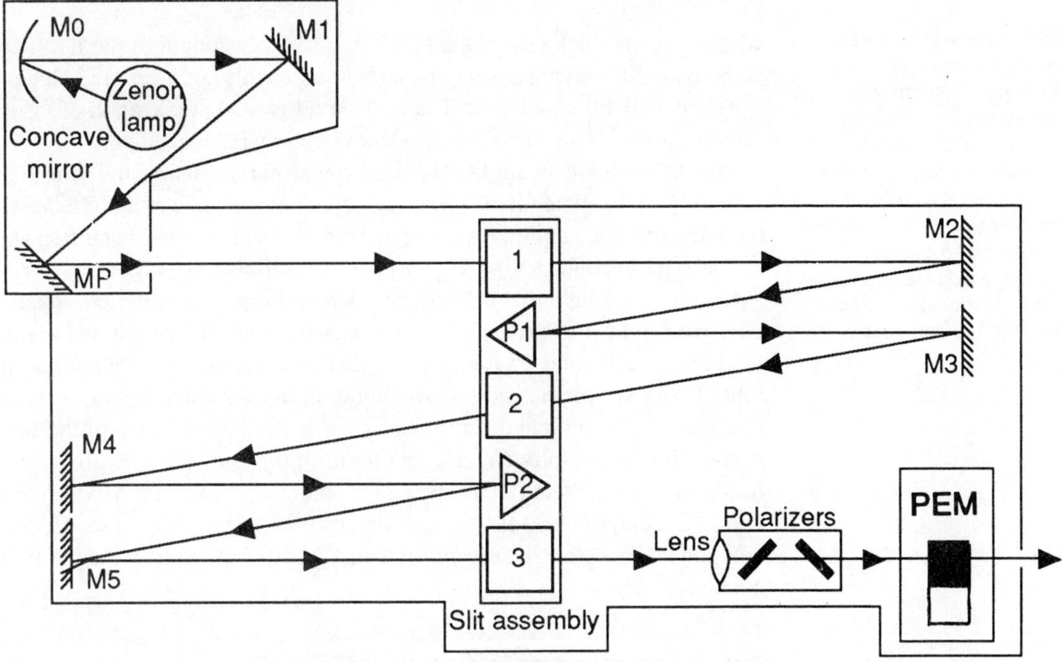

Fig. 7.2 Schematic diagram of the optics of a *CD* spectropolarimeter. In this instrument the monochromator prisms (P1 and P2) are also the polarizers; to further improve the polarization a series of tilted quartz plates are inserted in the light path preceding the PEM. Sample and photomultiplier tube are attached after the PEM. The xenon source means that the instrument has most sensitivity in the 300–400 nm region of the spectrum. The fact that a lamp is getting old becomes apparent by decreased light intensity in the 200 nm and 700 nm regions of the spectrum.

The optics of a typical *CD* machine are illustrated in Fig. 7.2. A series of mirrors and prisms and slits are used to produce collimated monochromatic radiation (in reality it is not monochromatic but has a well-defined wavelength range, typically of 0.5–2 nm, as discussed in §1.3). This light is then linearly polarized and subsequently circularly polarized. The conversion of linearly polarized light into circularly polarized light is achieved by the photo-elastic modulator, PEM.

The PEM consists of a piece of crystal quartz mechanically coupled (glued) to a piece of isotropic (silica) quartz, the light passing through the latter. The quartz plate is oriented so that light polarized (say) vertically travels through it at one speed and light polarized horizontally travels more slowly. If the linearly polarized light is incident on the plate at a 45° angle, then it may be considered as split into equal magnitude vertical and horizontal components. The PEM exhibits birefringence so that $(n_Z - n_Y) \neq 0$, *i.e.* the refractive indices n_Z and n_Y for vertical and horizontal polarizations are different. During passage through the PEM the phase difference between the vertical and horizontal light beam components will amount to:

$$\delta = \frac{2\pi D}{\lambda}$$
$$= \frac{2\pi d(n_Z - n_Y)}{\lambda} \tag{7.5}$$

where d is the thickness of the PEM, λ is the wavelength of the light, and D is the path difference between the light components. At a certain value of $(n_Z - n_Y)$, δ will be equal to $\pi/2$ and D equal to $\lambda/4$, making the so-called 'quarter-wave' plate required to produce circularly polarized light.

An AC voltage is applied to the crystal part, causing the whole PEM assembly to oscillate (at 50 kHz). By adjusting the voltage amplitude so that the birefringence amplitude corresponds to the quarter-wave condition at each λ, the time dependence of $(n_Z - n_Y)$ as it oscillates between $+|n_Z - n_Y|$ and $-|n_Z - n_Y|$ gives light that is alternately left and right circularly polarized. The polarized light thus produced then passes through the sample compartment (and the sample) and what is not absorbed is detected by the photomultiplier tube. If the sample has no *CD* the photomultiplier current will be constant (DC) with a magnitude determined by the normal absorbance of the sample. If the sample exhibits *CD* the photomultiplier current will also show an oscillating component (AC). The *CD* is obtained as the ratio between the AC and DC components. Its sign is determined from the phase of the AC component using a lock-in amplifier that has the AC voltage of the PEM as a time reference.

7.3 Interaction of radiation with matter

Using classical mechanics, the energy of the interaction, H^{int}, between the electric, **E**, and magnetic, **B**, fields of electromagnetic radiation and the molecule is given by

$$H^{int} = -\boldsymbol{\mu} \cdot \mathbf{E} - \boldsymbol{m} \cdot \mathbf{B} + \text{higher multipole terms} \tag{7.6}$$

where $\boldsymbol{\mu}$ is the electric dipole moment of the molecule and $\boldsymbol{m}$ is its magnetic dipole moment. $\boldsymbol{\mu}$ and $\boldsymbol{m}$ are determined by summing over the electron distribution of the molecule:

$$\boldsymbol{\mu} = e\sum_i (r_i)$$

$$\boldsymbol{m} = \frac{e}{2m_e}\sum_i \{r_i \times p_i\} \tag{7.7}$$

A factor of 2 increase in the voltage across the quarter-wave PEM that turns linearly polarized light into alternating left and right circularly polarized light for *CD* measurements, turns the PEM into a half-wave plate that produces alternating pulses of orthogonal beams of linearly polarized light for *LD* spectroscopy. An alternative that avoids the extra stress on the PEM is to insert an achromatic quarter-wave plate in the sample compartment.[3,4]

A brief discussion of vector products is given in Appendix 1.

$\boldsymbol{m}$ is a circulation of charge about an axis, hence the cross product (see Appendix 1) in eqn (7.7).

where e is the (negative) unit charge on an electron, r_i denotes the position vector of the ith electron, p_i is the momentum vector for the ith electron, and m_e is the mass of an electron.

When we are interested in spectroscopic phenomena such as absorbance, CD, and LD, we have to move beyond classical mechanics to quantum mechanics since the transitions we study involve discrete amounts of energy —*i.e.* the energy is quantized. Thus, instead of an expression for the energy of the interaction between the radiation and the molecule, we need the interaction Hamiltonian. This follows from eqn (7.6) by converting μ and m to operators and the interaction energy to the Hamiltonian operator. This is denoted by using curved 'hats' to indicate operators:

$$\widehat{H}^{\text{int}} = -\widehat{\mu} \cdot \mathbf{E} - \widehat{m} \cdot \mathbf{B} + \text{higher order terms} \tag{7.8}$$

When we compare the relative strengths of different transitions experimentally we measure the intensity of the transition which is the area under the absorption band. The related theoretical quantity is the probability per unit time, P, that a transition will occur from the initial state, $|i\rangle$, (which is usually the ground state) to the final state, $|f\rangle$:[1,5]

$$P(|i\rangle \to |f\rangle) = k' \left|\langle f|\widehat{H}^{\text{int}}|i\rangle\right|^2$$
$$= k' \langle f|\widehat{H}^{\text{int}}|i\rangle^* \langle f|\widehat{H}^{\text{int}}|i\rangle \tag{7.9}$$
$$= k' \langle i|\widehat{H}^{\text{int}}|f\rangle\langle f|\widehat{H}^{\text{int}}|i\rangle$$

where the integral is over molecular position coordinates, and k' is a constant. Equation (7.9) is the Fermi Golden Rule. Upon substituting the expression for $\widehat{H}^{\text{int}}$ into eqn (7.9) we have

$$P(|i\rangle \to |f\rangle) = k\left\{\langle i|\widehat{\mu}|f\rangle \cdot \widehat{\mathbf{e}}^* + \langle i|\widehat{m}|f\rangle \cdot \widehat{\mathbf{b}}^*\right\}\left\{\langle f|\widehat{\mu}|i\rangle \cdot \widehat{\mathbf{e}} + \langle f|\widehat{m}|i\rangle \cdot \widehat{\mathbf{b}}\right\} \tag{7.10}$$

where $\widehat{\mathbf{b}}$ is the unit vector along $\mathbf{B}$, the asterisk denotes the complex conjugate, $|i\rangle^* = \langle i|$, and the magnitudes of the electric and magnetic fields are absorbed into k. $\langle f|\widehat{\mu}|i\rangle$ is the edtm *from* state i *to* state f, and $\langle f|\widehat{m}|i\rangle$ the corresponding mdtm. It follows that

$$P(|i\rangle \to |f\rangle) = k\left[\mu^{fi} \cdot \widehat{\mathbf{e}}^*\mu^{if} \cdot \widehat{\mathbf{e}} + m^{fi} \cdot \widehat{\mathbf{b}}^* m^{if} \cdot \widehat{\mathbf{b}}\right]_{(I)}$$
$$+ k\left[\mu^{if} \cdot \widehat{\mathbf{e}}\, m^{fi} \cdot \widehat{\mathbf{b}}^* + \mu^{fi} \cdot \widehat{\mathbf{e}}^* m^{if} \cdot \widehat{\mathbf{b}}\right]_{(II)} \tag{7.11}$$

where $\mu^{if} = \mu^{i\to f} = \langle f|\widehat{\mu}|i\rangle$ *etc.* (note inversion of order of states in the notation).

In the next subsections we consider a number of particular cases where a molecule interacts with radiation of different polarizations.

Interaction of a molecule with linearly polarized light: A and LD

If no external magnetic field is present (except that due to the radiation) we can choose $|i\rangle$ and $|f\rangle$ to be real wavefunctions; thus $\mu^{fi} = \mu^{if}$ but $m^{fi} = -m^{if}$ as $\widehat{m}$ is imaginary. The polarization vectors for $\mathbf{E}$, and hence $\mathbf{B}$, are also real for linearly polarized light. It follows that the two parts of term

When we use what is known as bra–ket notation, the 'bra' $\langle 0|$ state is the complex conjugate of the 'ket' $|0\rangle$, and when they are coupled together either directly or either side of an operator, such as $\widehat{\mu}$ in eqn (7.10), then we imply that an integral over all space is being performed. We shall use round brackets $|\ )$ for the unperturbed states of chromophores $\mathbf{A}$ and $\mathbf{C}$ (see below), and pointed ones $|\ \rangle$ for the states of the whole system.

See Appendix 4 for the relationships between these theoretical equations and the experimental quantities.

(*II*) in eqn (7.11) are equal in magnitude but opposite in sign, so term *II* is zero. As magnetic terms are invariably small compared with electric ones, we can ignore the second term within term (*I*) and write the probability of the transition occurring from $|i\rangle$ to $|f\rangle$ with light polarized parallel to the electric field (the absorbance, A_e, of the transition) to be:

$$A_e\left(|i\rangle \rightarrow |f\rangle\right) = k\left|\boldsymbol{\mu}^{if} \cdot \hat{\mathbf{e}}\right|^2 \tag{7.12}$$

Isotropic absorption of linearly polarized light
For the usual situation where the system is a collection of randomly oriented randomly placed molecules, eqn (7.12) becomes:

$$A_{iso} = A = \frac{k}{3}\left|\mu^{if}\right|^2 \tag{7.13}$$

Linear dichroism
Linear dichroism (*LD*) was defined in Chapter 1 to be the difference in absorption of two in-phase linearly polarized light beams propagating in the same direction with perpendicular polarizations. Take the ∥ direction to be (0,0,1) and the perpendicular direction to be (0,1,0). Then from eqn (7.12)

$$LD_{molecule} = (A_\parallel - A_\perp)_{molecule}$$
$$= k\left\{\left|\mu_Z^{if}\right|^2 - \left|\mu_Y^{if}\right|^2\right\} \tag{7.14}$$

where μ_Y^{if} is the Y component of the $|i\rangle$ to $|f\rangle$ transition dipole moment, *etc.* For a collection of N molecules we sum over the *LD* for each molecule

$$LD_{total} = \sum_{molecules}(A_\parallel - A_\perp)_{molecule}$$
$$= Nk\left\{\left\langle\left|\mu_Z^{if}\right|^2\right\rangle - \left\langle\left|\mu_Y^{if}\right|^2\right\rangle\right\} \tag{7.15}$$

where $\langle\ \rangle$ denotes average. The net *LD* thus depends on the method and extent of sample orientation since $\{X,Y,Z\}$ is a laboratory-fixed axis system about which the molecule knows nothing unless we tell it. In the absence of any molecular orientation the *LD* in eqn (7.15) vanishes because

$$\left\langle\left|\mu_Z^{if}\right|^2\right\rangle = \left\langle\left|\mu_Y^{if}\right|^2\right\rangle = \frac{1}{3}\left|\mu^{if}\right|^2 \tag{7.16}$$

Interaction of a molecule with circularly polarized light: *CD*
As *CD* is the difference in absorption of left and right circularly polarized light by a molecule, as long as there is no magnetic field apart from that of the radiation present, we can again choose to use real wavefunctions so $\boldsymbol{\mu}^{fi} = \boldsymbol{\mu}^{if}$ and $\boldsymbol{m}^{fi} = -\boldsymbol{m}^{if}$. To evaluate eqn (1.1) we substitute first the expression for left circularly polarized light into eqn (7.11) then subtract from it the result of substituting the expression for right circularly polarized light. We also make use of the equality $\hat{\mathbf{e}}_r = \hat{\mathbf{e}}_\ell^*$, and note that (§1.2)

$$\hat{\mathbf{b}} = \mathbf{k} \times \hat{\mathbf{e}} \tag{7.17}$$

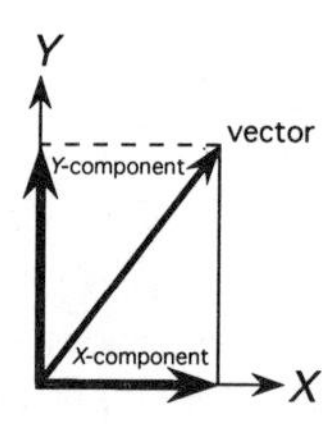

so

$$\hat{\mathbf{b}}_\ell = \frac{1}{\sqrt{2}}(1,0,0)\times(0,1,\mathrm{i})$$

$$= \frac{1}{\sqrt{2}}(0,-\mathrm{i},1) = \hat{\mathbf{b}}_r^* \qquad (7.18)$$

Thus

$$CD = A_\ell - A_r$$

$$= k\left[\begin{array}{c}\left(\boldsymbol{\mu}^{fi}\cdot\hat{\mathbf{e}}_\ell^*\boldsymbol{\mu}^{if}\cdot\hat{\mathbf{e}}_\ell + \mathbf{m}^{fi}\cdot\hat{\mathbf{b}}_\ell^*\mathbf{m}^{if}\cdot\hat{\mathbf{b}}_\ell\right)\\ -\left(\boldsymbol{\mu}^{fi}\cdot\hat{\mathbf{e}}_\ell\boldsymbol{\mu}^{if}\cdot\hat{\mathbf{e}}_\ell^* + \mathbf{m}^{fi}\cdot\hat{\mathbf{b}}_\ell\mathbf{m}^{if}\cdot\hat{\mathbf{b}}_\ell^*\right)\end{array}\right]_{(I)} \qquad (7.19)$$

$$+\, k\left[\begin{array}{c}\left(\boldsymbol{\mu}^{if}\cdot\hat{\mathbf{e}}_\ell\,\mathbf{m}^{fi}\cdot\hat{\mathbf{b}}_\ell^* + \boldsymbol{\mu}^{fi}\cdot\hat{\mathbf{e}}_\ell^*\mathbf{m}^{if}\cdot\hat{\mathbf{b}}_\ell\right)\\ -\left(\boldsymbol{\mu}^{if}\cdot\hat{\mathbf{e}}_\ell^*\,\mathbf{m}^{fi}\cdot\hat{\mathbf{b}}_\ell + \boldsymbol{\mu}^{fi}\cdot\hat{\mathbf{e}}_\ell\mathbf{m}^{if}\cdot\hat{\mathbf{b}}_\ell^*\right)\end{array}\right]_{(II)}$$

Because $\boldsymbol{\mu}^{fi}\cdot\hat{\mathbf{e}}_\ell^*\boldsymbol{\mu}^{if}\cdot\hat{\mathbf{e}}_\ell = \boldsymbol{\mu}^{fi}\cdot\hat{\mathbf{e}}_\ell\boldsymbol{\mu}^{if}\cdot\hat{\mathbf{e}}_\ell^*$ *etc.* (the wavefunctions are real), the first term (*I*) vanishes. Upon rearranging term (*II*) so that the electric dipole transition moments are all from $|f\rangle$ to $|i\rangle$ and the magnetic moments are the reverse, eqn (7.19) then becomes

$$CD = A_\ell - A_r$$

$$= 2k\left[\boldsymbol{\mu}^{fi}\cdot\hat{\mathbf{e}}_\ell^*\,\mathbf{m}^{if}\cdot\hat{\mathbf{b}}_\ell - \boldsymbol{\mu}^{fi}\cdot\hat{\mathbf{e}}_\ell\,\mathbf{m}^{if}\cdot\hat{\mathbf{b}}_\ell^*\right] \qquad (7.20)$$

Upon substituting explicit forms for the electric [eqn (7.4)] and magnetic field polarizations for the circularly polarized radiation we get

$$CD = -2\mathrm{i}k\left[\mu_Y^{fi}m_Y^{if} + \mu_Z^{fi}m_Z^{if}\right]$$

$$= 2k\,\mathrm{Im}\left[\mu_Y^{fi}m_Y^{if} + \mu_Z^{fi}m_Z^{if}\right] \qquad (7.21)$$

The second line of eqn (7.21) is also true for complex wavefunctions.

where 'Im' denotes imaginary part of.

CD of a collection of randomly oriented molecules

CD experiments are usually performed on collections of randomly oriented molecules. In that case the laboratory (radiation)-defined axis system does not relate in any fixed way to a molecular axis system, so what is measured is a rotational average of eqn (7.21):

$$CD = \frac{4}{3}k\langle i|\tilde{\boldsymbol{\mu}}|f\rangle\cdot\langle f|\tilde{\mathbf{m}}|i\rangle = \frac{4}{3}k\,\mathrm{Im}\left[\boldsymbol{\mu}^{fi}\cdot\mathbf{m}^{if}\right] \qquad (7.22)$$

The factor of 2/3 arises because the rotational average of each term in eqn. (7.21) gives rise to 1/3.

which is the Rosenfeld equation [eqn (5.1)] if we ignore the constants.

Interaction of a molecule with circularly polarized light in the presence of an external static magnetic field: *MCD*

In §6.6 we looked at the *CD* signal, the so-called magnetic *CD*, that can be measured for achiral molecules when an external static magnetic field gives eda transitions some helical character. In deriving the equations for such a situation where an external magnetic field is imposed, the wavefunctions can no longer be taken to be real. Thus $\boldsymbol{\mu}^{fi}\cdot\hat{\mathbf{e}}_\ell^*\boldsymbol{\mu}^{if}\cdot\hat{\mathbf{e}}_\ell$ may differ from $\boldsymbol{\mu}^{fi}\cdot\hat{\mathbf{e}}_\ell\boldsymbol{\mu}^{if}\cdot\hat{\mathbf{e}}_\ell^*$ and term (*I*) in eqn (7.19) does not vanish. However, if, as in

Chapter 6, we limit our consideration to achiral molecules then term (*II*) does vanish. Further, we ignore the terms that involve squares of mdtms as they will be much smaller than any etdm terms. The magnetic *CD* (*MCD*) may then be written

General expressions for *MCD* often have quadrupole terms, which come from higher order terms in the interaction operator.

$$MCD = k\left[\left(\mu_Y^{fi} - i\mu_Z^{fi}\right)\left(\mu_Y^{if} + i\mu_Z^{if}\right) - \left(\mu_Y^{fi} + i\mu_Z^{fi}\right)\left(\mu_Y^{if} + i\mu_Z^{if}\right)\right]$$

$$= 2ik\left(\mu_Y^{fi}\mu_Z^{if} - \mu_Z^{fi}\mu_Y^{if}\right) \tag{7.23}$$

In §7.9 we shall use eqn (7.23) to derive the equations used in §6.6 and see how to determine spectroscopic and geometric information about a molecule from *MCD* spectra.

7.4 *CD*, transition moment operators, and transition moments

What happens when we measure a *CD* spectrum is summarized in Fig. 7.3. In this section we shall see how to express transition moments of a system in terms of the moments of its (usually achiral) sub-units. In developing the equations for the *CD* resulting from the coupling of independent chromophores in this and the succeeding sections of the chapter we shall begin by using perturbation theory to write expressions for the ground and excited state wavefunctions between which the transition of interest takes place.

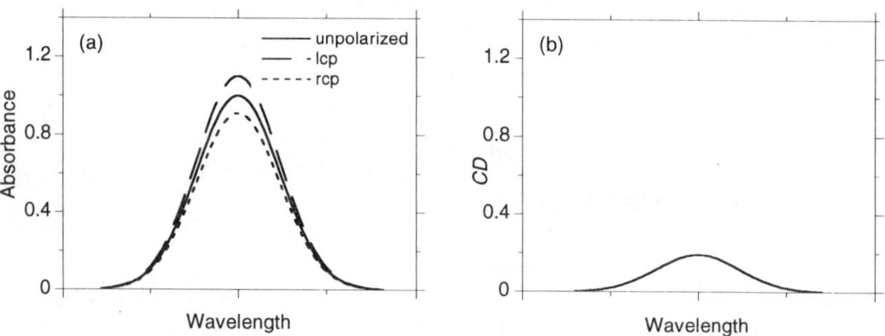

Fig. 7.3 (a) Spectra from absorption of unpolarized, left circularly polarized, and right circularly polarized light by a randomly oriented sample of chiral molecules. (b) The resulting *CD* spectrum. The differences in absorption have been exaggerated for illustrative purposes.

The second stage is to substitute those expressions into the Rosenfeld equation for the rotatory or *CD* strength

$$R = CD = \text{Im}\langle 0|\hat{\boldsymbol{\mu}}|1\rangle \cdot \langle 1|\hat{\boldsymbol{m}}|0\rangle = \text{Im}\left(\boldsymbol{\mu}^{10} \cdot \boldsymbol{m}^{01}\right) \tag{7.24}$$

where we use the notation defined above but use $|0\rangle$ and $|1\rangle$ rather than $|i\rangle$ and $|f\rangle$ for the initial and final state of the transition. $\boldsymbol{\mu}^{10}$ is thus the electric dipole transition moment (edtm) for the transition *from* the excited *to* the ground state and $\boldsymbol{m}^{01}$ is the magnetic dipole transition moment (mdtm) for the reverse transition.

In the remainder of this chapter we shall be working within the chromophoric approach to *CD* theory (§5.3), and will limit our consideration to transitions in achiral chromophores. A *chromophore* is strictly a part of a system whose wavefunctions have *no* overlap with the rest of the system; electronic wavefunctions in a chromophore therefore have no electron exchange with the rest of the system. In practice, a chromophore is usually identified for a given electronic transition if that transition seems to be more or less dependent on only the identity of a subset of the molecule (such as the carbonyl functional group).

The Rosenfeld equation for a transition within an isolated *achiral* chromophore **A** gives no *CD*:

$$CD = \mathrm{Im}\big(0|\hat{\boldsymbol{\mu}}|1\big)\cdot\big(1|\hat{m}|0\big) = \mathrm{Im}\big(\boldsymbol{\mu}_A^{10}\cdot\boldsymbol{m}_A^{01}\big) = 0 \qquad (7.25)$$

where we use curved bra-ket notation for the states of isolated (usually achiral) chromophores and subscripts on the transition moments to denote that they belong entirely to the chromophore **A**. Eqn (7.25) vanishes because the electron displacement of a transition within an achiral chromophore is not helical, it is either linear ($m_A^{01} = 0$), or circular ($\boldsymbol{\mu}_A^{01} = 0$), or a planar spiral if the two moments are perpendicular (§5.2).

As soon as **A** is allowed to interact with other chromophores, $\{\mathbf{C}_i\}$, unless the whole system is still achiral we should expect to be able to see a *CD* signal induced into the **A** transitions. Since the *CD* strength depends crucially on the geometry of the whole system, the *CD* spectrum *should* be able to tell us about the arrangement of the $\mathbf{C}_i$ about **A**. The way we extract that information is different depending whether the **A** transition is electric dipole allowed (eda) or electric dipole forbidden (edf) but magnetic dipole allowed (mda). For the rest of this chapter we shall generally consider two chromophores, **A** and **C**. Including more $\mathbf{C}_i$ simply requires the inclusion of a sum over i.

To evaluate the Rosenfeld equation for two interacting chromophores we need to know $\boldsymbol{\mu}^{01}$ and $\boldsymbol{m}^{01}$ for the whole system, *i.e.* for the $|0\rangle \to |1\rangle$ transition, which is the $|0) \to |1)$ transition on **A** in the presence of any perturbation due to **C**. We can write the wavefunctions for the **A**/**C** system when their interaction is switched off as the product wavefunctions $|a)|c) = |ac)$ where $|a)$ is a wavefunction of **A** and $|c)$ is a wavefunction of **C**. We always write the **A** function first, even in the ket form, thus

$$\big\{|a)|c)\big\}^{*} = \big\{|ac)\big\}^{*} = (a|(c| = (ac| \qquad (7.26)$$

A operators only operate on **A** wavefunctions; similarly **C** operators only operate on **C** wavefunction. For example,

$$(a'\,c'|\hat{\boldsymbol{\mu}}_A|ac) = (a'|\hat{\boldsymbol{\mu}}_A|a)(c'|c)$$
$$= \mu_A^{aa'}\delta_{cc'} \qquad (7.27)$$

where $\delta_{cc'} = 0$ unless $c' = c$.

$(c'|c)$ is the integral over **C** coordinates of the product of the wavefunctions $|c')^{*}$ and $|c)$. It vanishes unless $c' = c$, in which case, since we have usually chosen the functions to be normalized, the integral equals 1. This is denoted by the Kronecker delta symbol $\delta_{cc'}$ which equals 1 if $c' = c$, and equals zero otherwise.

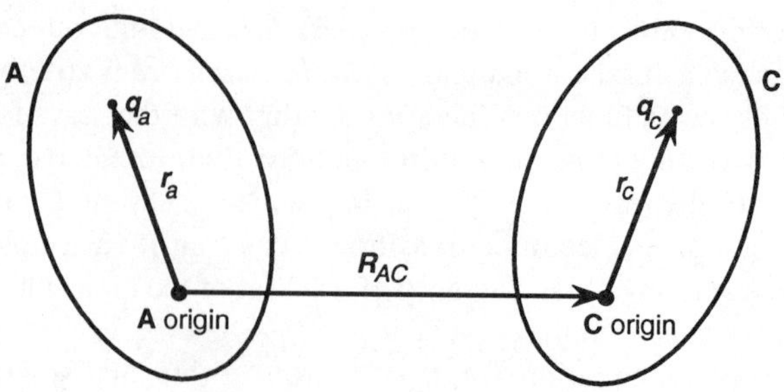

Fig. 7.4 A and **C** chromophores.

We also need to be able to write an equation for the interaction between **A** and **C**. Since there is no electron exchange between **A** and **C** (by definition), their interaction, V, is purely coulombic

$$V = \sum_{a,c} \frac{q_a q_c}{\left| \boldsymbol{R}_{AC} + \boldsymbol{r}_c - \boldsymbol{r}_a \right|} \tag{7.28}$$

where q_a is the charge of a particle in **A** located at the end of the vector, $\boldsymbol{r}_a$, which begins at the **A** origin, similarly q_c and $\boldsymbol{r}_c$, and $\boldsymbol{R}_{AC} = \boldsymbol{R}_C - \boldsymbol{R}_A$ is the vector from the origin of chromophore **A** to the origin of chromophore **C** (Fig. 7.4).

Now, from eqn (7.7) the electric dipole moment operator, $\tilde{\boldsymbol{\mu}}$, where $\tilde{\ }$ is used to denote the operators for the whole system is

$$\tilde{\boldsymbol{\mu}} = e \sum_i \left(\hat{\boldsymbol{r}}_i \right) \tag{7.29}$$

where $\hat{\boldsymbol{r}}_i$ is the position operator in the global coordinate system for a particle in **A** or **C**. We may rewrite eqn (7.29) in terms of **A** and **C** operators as

$$\tilde{\boldsymbol{\mu}} = e \left\{ \sum_a \left(\hat{\boldsymbol{R}}_A + \hat{\boldsymbol{r}}_a \right) + \sum_c \left(\hat{\boldsymbol{R}}_C + \hat{\boldsymbol{r}}_c \right) \right\} \tag{7.30}$$

where $\hat{\boldsymbol{R}}_A = \boldsymbol{R}_A$ is the position operator for the **A** origin in the global coordinate system, and $\hat{\boldsymbol{r}}_a = \boldsymbol{r}_a$ is the position operator within **A** for particle a (Fig. 7.4). We shall be considering only *transition* moments, so since $\boldsymbol{R}_A$ is constant it follows that

$$\tilde{\boldsymbol{\mu}} = e \left\{ \sum_a \left(\hat{\boldsymbol{r}}_a \right) + \sum_c \left(\hat{\boldsymbol{r}}_c \right) \right\}$$
$$= \hat{\boldsymbol{\mu}}_A + \hat{\boldsymbol{\mu}}_C \tag{7.31}$$

where $\hat{\boldsymbol{\mu}}_A$ is the electric dipole moment operator for **A** and operates only on wavefunctions of **A**.

Similarly it follows from eqn (7.7) that the magnetic dipole moment operator for the whole system is

$$\tilde{\boldsymbol{m}} = \frac{e}{2m_e} \sum_i \left\{ \hat{\boldsymbol{r}}_i \times \hat{\boldsymbol{p}}_i \right\} \tag{7.32}$$

Eqn (7.32) may be rewritten

We shall usually end up defining the system origin to be the **A** chromophore origin so

$$\hat{\boldsymbol{R}}_A = \boldsymbol{R}_A = (0,0,0)$$

and

$$\boldsymbol{R}_{AC} = \boldsymbol{R}_C - \boldsymbol{R}_A = \boldsymbol{R}_C$$

If Θ is a constant, then

$$\langle k | \hat{\Theta} | j \rangle = \Theta \langle k | j \rangle = \Theta \delta_{jk}$$

where $\delta_{jk} = 0$ since $j \neq k$ for transitions between states.

$$\tilde{m} = \tilde{m}_A + \tilde{m}_C$$

$$= \frac{e}{2m_e}\left\{\sum_a\left(\hat{R}_A + \hat{r}_a\right)\times\hat{p}_a + \sum_c\left(\hat{R}_C + \hat{r}_c\right)\times\hat{p}_c\right\}$$

$$= \frac{e}{2m_e}\hat{R}_A\times\sum_a\hat{p}_a + \hat{m}_A + \frac{e}{2m_e}\hat{R}_C\times\sum_c\hat{p}_c + \hat{m}_C \qquad (7.33)$$

$$= \frac{e}{2m_e}\hat{R}_A\times\hat{p}_A + \hat{m}_A + \frac{e}{2m_e}\hat{R}_C\times\hat{p}_C + \hat{m}_C$$

where $\tilde{m}_A$ is the total magnetic moment for chromophore **A**, $\hat{p}_A$ is the linear momentum operator for chromophore **A**, $\hat{m}_A$ is the intrinsic magnetic dipole moment operator within **A** *etc.*

When we come to evaluate any *CD* expression, the operators are always 'sandwiched' between two different states, $|k)$ of energy ε_k and $|j)$ of energy ε_j, just as in the transition moment integrals. In fact, it is convenient to re-express transition momenta in terms of electric dipole transition moments for the appropriate transition exploiting a very useful relationship that is derived in Appendix 3:

$$\left(k|\hat{p}_C|j\right) = \frac{im_e}{e\hbar}\left(\varepsilon_k - \varepsilon_j\right)\boldsymbol{\mu}_C^{jk} \qquad (7.34)$$

from which, if there is no intrinsic magnetic moment for the transition within **C** (*i.e.* it is eda and mdf) it follows

$$m_C^{jk} = \left(k|\tilde{m}_C|j\right)$$

$$= \frac{e}{2m_e}\left(k|\hat{R}_C\times\hat{p}_C|j\right) \qquad (7.35)$$

$$= \frac{i}{2\hbar}R_C\times\left(\varepsilon_k - \varepsilon_j\right)\boldsymbol{\mu}_C^{jk}$$

This is the magnetic moment at the system origin created by the tangential eda transition moment $\boldsymbol{\mu}_C^{jk}$.

7.5 *CD* from the coupling of degenerate electric dipole transition moments in identical distinct chromophores: the degenerate coupled-oscillator model

If an eda transition, with edtm $\boldsymbol{\mu}_A^{01}$ and transition energy ε, of an *achiral* chromophore, **A**, has a measurable *CD* spectrum, then it must have 'borrowed' some magnetic character, to give a net helical electron rearrangement. The degenerate coupled-oscillator model describes the situation where the magnetic character comes from the coupling of $\boldsymbol{\mu}_A^{01}$ with another edtm in **C**, $\boldsymbol{\mu}_C^{01}$, whose transition energy is the same but whose orientation is skewed relative to that of $\boldsymbol{\mu}_A^{01}$. $\boldsymbol{\mu}_C^{01}$ induces a magnetic effect in **A** since, although it is a linear motion of charge within **C**, when viewed from **A** it moves around **A**'s origin as illustrated in Fig. 5.4 ($\boldsymbol{\mu}_A^{01}$ simultaneously induces a magnetic component into the **C** transition $\boldsymbol{\mu}_C^{01}$).

To write the wavefunctions of the interacting **A/C** system we first consider an **A/C** system where **A** and **C** do not interact. The Hamiltonian, H, of the non-interacting system is then simply the sum

$$H = H_A + H_C \qquad (7.36)$$

The total magnetic moment for **A** includes both the intrinsic magnetic moment operator within the chromophore and any magnetic moment due to the linear motion of its charges about the system origin.

where H_A is the Hamiltonian of isolated **A**, and similarly H_C. The total energy of a state $|ac)$ is the energy of $|a)$ in **A** plus that of $|c)$ in **C**. So, for example, $|01)$ means **A** is in its ground state and **C** is in state $|1)$. The energy of the $|01)$ state of the combined system is therefore $0 + \varepsilon = \varepsilon$ (taking the zero point of energy to be when both **A** and **C** are in their ground states).

When the electrostatic interaction, V [eqn (7.28)], is 'switched' on between **A** and **C** then the Hamiltonian of the interacting system is

$$H = H_A + H_C + V \tag{7.37}$$

If **A** and **C** are not too close together then V may be expanded using a double Taylor series expansion to give terms dependent on monopoles, dipoles, quadrupoles *etc.* of **A** and **C**. For uncharged **A** and **C**, the first term in the expansion is the dipole–dipole term:

$$V = \frac{\boldsymbol{\mu}_A \cdot \boldsymbol{\mu}_C - 3\hat{\boldsymbol{R}}_{AC} \cdot \boldsymbol{\mu}_A \boldsymbol{\mu}_C \cdot \hat{\boldsymbol{R}}_{AC}}{R_{AC}^3} \tag{7.38}$$

where $\boldsymbol{R}_{AC}$ is the vector from the **A** origin to the **C** origin. The **A/C** wavefunctions have the form

$$|a)|c) = |ac\rangle$$
$$= |ac) + \textit{other terms} \tag{7.39}$$

The identity of the 'other terms' depends on whether or not **A** and **C** are identical. If **A** and **C** are identical then their transitions are degenerate (have the same energy) and by symmetry (or degenerate perturbation theory) the two states that result from the mixing of $|10)$ (**A** in $|1)$ and **C** in the ground state) and $|01)$ are

$$|10_\pm\rangle = \frac{1}{\sqrt{2}}\left\{|10) \pm |01)\right\} \tag{7.40}$$

As above, we assume real normalized wavefunctions. If we also assume that permanent moments are much smaller than transition moments, then the energies of these two states are

$$\varepsilon^\pm = \left\langle 10_\pm \left| H_A + H_C \pm V \right| 10_\pm \right\rangle$$
$$= \tfrac{1}{2}\left\{ (10| \pm (01| \right\} | H_A + H_C + V | \left\{ |10) \pm |01) \right\}$$
$$= \tfrac{1}{2}\left\{ (1|H_A|1) + (0|H_C|0) \pm (10|V|01) + (0|H_A|0) + (1|H_C|1) \pm (01|V|10) \right\}$$
$$= \varepsilon^0 + \varepsilon^1 \pm V^{10,01}$$
$$= \varepsilon \pm V^{11} \tag{7.41}$$

where, for example,

$$V^{ba,dc} = (ac|V|bd)$$
$$= \frac{\boldsymbol{\mu}_A^{ba} \cdot \boldsymbol{\mu}_C^{dc} - 3\hat{\boldsymbol{R}}_{AC} \cdot \boldsymbol{\mu}_A^{ba} \, \boldsymbol{\mu}_C^{dc} \cdot \hat{\boldsymbol{R}}_{AC}}{R_{AC}^3} \tag{7.42}$$

Because all wavefunctions are real, when one state on each of **A** and **C** is the ground state we further simplify the notation as indicated in the last line of eqn (7.41). V^{11} is much smaller than ε so the two new perturbed states are close in energy.

The edtms for the transitions from the two excited states of eqn (7.40) to the ground state $|00\rangle$ are:

$$\langle 00|\tilde{\boldsymbol{\mu}}|10_{\pm}\rangle = \frac{1}{\sqrt{2}}\left\{(00|\hat{\boldsymbol{\mu}}_A + \hat{\boldsymbol{\mu}}_C[|10\rangle \pm |01\rangle)]\right\}$$

$$= \frac{1}{\sqrt{2}}\left\{\boldsymbol{\mu}_A^{10} \pm \boldsymbol{\mu}_C^{10}\right\} \tag{7.43}$$

where the upper sign of the $\pm$ refers to the upper sign in eqns (7.40) and (7.41).

The $|0\rangle \rightarrow |1\rangle$ transitions in both **A** and **C** are mdf, and so have no intrinsic magnetic moment. We therefore set $\hat{\boldsymbol{m}}_A = \hat{\boldsymbol{m}}_C = 0$ in eqn (7.33) and write the mdtms for the transition from the ground states to the excited perturbed states to be

$$\left(10_{\pm}|\tilde{\boldsymbol{m}}|00\right) = \frac{i\varepsilon}{\sqrt{8}\hbar}\left\{\boldsymbol{R}_A \times \boldsymbol{\mu}_A^{01} \pm \boldsymbol{R}_C \times \boldsymbol{\mu}_C^{01}\right\} \tag{7.44}$$

Substituting eqns (7.43) and (7.44) into the Rosenfeld equation gives the *CD* strengths of bands centred at ε^+ and ε^- [as given by eqn (7.41)] arising from the coupling of degenerate $|0\rangle \rightarrow |1\rangle$ transitions on **A** and **C** to be

$$R_{\pm}^{01} = \pm\frac{\varepsilon}{4\hbar}\left\{\boldsymbol{\mu}_C^{01} \times \boldsymbol{\mu}_A^{01} \cdot \left(\boldsymbol{R}_C - \boldsymbol{R}_A\right)\right\}$$

$$= \pm\frac{\varepsilon}{4\hbar}\left\{\boldsymbol{\mu}_C^{01} \times \boldsymbol{\mu}_A^{01} \cdot \boldsymbol{R}_{AC}\right\} \tag{7.45}$$

where the $\pm$ subscripts refer to the '+' and '−' states respectively in eqn (7.40).

Eqn (5.8) is a notationally simplified version of eqn (7.45). The physical interpretation and applications of eqn (7.45) for solving geometric and spectroscopic problems is outlined in Chapter 5.

If there are $n > 2$ identical chromophores in the system arranged around an n-fold rotational symmetry axis then eqn (7.40) is replaced by the two symmetry adapted projections and normalization factors. The result is always a transition polarized along the n-fold axis, z, whose *CD* is always equal in magnitude but opposite in sign from a second transition degenerately polarized in the plane perpendicular to the n-fold axis. The normalization factor for the z-polarized transition is $1/\sqrt{n}$. Assuming that each chromophore couples only with its two nearest neighbours, the *CD* signals are then[6]

$$R(z, n \text{ chromophores}) = \left(\frac{\sqrt{2}}{\sqrt{n}}\right)^2 nR_+(2 \text{ chromophores}) \tag{7.46}$$

$$= -R(x/y, n \text{ chromophores})$$

7.6 *CD* from the coupling of electric dipole transition moments in non-identical chromophores: the non-degenerate coupled-oscillator model

If **A** and **C** are not identical then the wavefunctions of the combined system are not determined by symmetry. This makes writing an expression for the wavefunctions and hence the transition moments more difficult, but the compensating factor is that because the **A** and **C** transitions occur at very different energies we can take the energies of the states of the interacting **A/C** system to be simple sums of the energies of the non-interacting **A** and **C** states.

Further, we need only determine the *CD* induced into the $|0)_A \rightarrow |1)_A$ transition at energy ε_a of **A**. So we require expressions for the perturbed states of the interacting **A/C** system where both chromophores are in the ground state, $|00\rangle$, or where **A** is in an excited state and **C** is in its ground state, $|10\rangle$. For simplicity, we shall consider only these two states and the two others $|0c\rangle$ and $|1c\rangle$ where **C** is in the excited state $|c\rangle$ on **C** (the final equations are easily generalized by introducing summations at the end). From perturbation theory we may write

$$|00\rangle = |00\rangle - \frac{(1c|V|00)}{\varepsilon_c + \varepsilon_a}|1c) \tag{7.47}$$

$$|10\rangle = |10\rangle - \frac{(0c|V|10)}{\varepsilon_c - \varepsilon_a}|0c) \tag{7.48}$$

It then follows (again assuming terms containing permanent moments are smaller than those containing transition moments) that to first order in V for

$$V^{1c} = \frac{\boldsymbol{\mu}_A^{01} \cdot \boldsymbol{\mu}_C^{c0} - 3\hat{\boldsymbol{R}}_{AC} \cdot \boldsymbol{\mu}_A^{01}\, \boldsymbol{\mu}_C^{c0} \cdot \hat{\boldsymbol{R}}_{AC}}{R_{AC}^3} \tag{7.49}$$

the edtm of the perturbed $|10\rangle \rightarrow |00\rangle$ transition is

$$\boldsymbol{\mu}^{10} = \langle 00|\tilde{\boldsymbol{\mu}}|10\rangle$$

$$= \langle 00|\hat{\boldsymbol{\mu}}_A + \hat{\boldsymbol{\mu}}_C|10\rangle$$

$$= (00|\hat{\boldsymbol{\mu}}_A|10) - (00|\hat{\boldsymbol{\mu}}_C|0c)\frac{(0c|V|10)}{\varepsilon_c - \varepsilon_a} - (1c|\hat{\boldsymbol{\mu}}_C|10)\frac{(00|V|1c)}{\varepsilon_c + \varepsilon_a} \tag{7.50}$$

$$= \boldsymbol{\mu}_A^{10} - \frac{2\varepsilon_c V^{1c}}{\varepsilon_c^2 - \varepsilon_a^2}\boldsymbol{\mu}_C^{c0}$$

Similarly,

$$\mathbf{m}^{01} = \langle 10|\tilde{\mathbf{m}}|00\rangle$$

$$= \frac{e}{2m_e}\left\{ \boldsymbol{R}_A \times \boldsymbol{p}_A^{01} - \frac{2\varepsilon_a V^{1c}}{\varepsilon_c^2 - \varepsilon_a^2}\boldsymbol{R}_C \times \boldsymbol{p}_C^{0c} \right\} \tag{7.51}$$

$$= \frac{\mathrm{i}}{2\hbar}\left\{ \varepsilon_a \boldsymbol{R}_A \times \boldsymbol{\mu}_A^{01} - \frac{2\varepsilon_a\varepsilon_c V^{1c}}{\varepsilon_c^2 - \varepsilon_a^2}\boldsymbol{R}_C \times \boldsymbol{\mu}_C^{0c} \right\}$$

since the transition has no intrinsic magnetic moment within **A**.

A little imprecision about the ordering of states with *electric* transition moments does not matter as long as the wavefunctions are real. The same is not true for *magnetic* dipole transition moments.

The *CD* strength is therefore

$$R(|00\rangle \to |10\rangle) = \frac{-\varepsilon_a \varepsilon_c V^{1c}}{\hbar\left(\varepsilon_c^2 - \varepsilon_a^2\right)}\left\{\boldsymbol{\mu}_A^{10} \cdot \boldsymbol{R}_C \times \boldsymbol{\mu}_C^{0c} + \boldsymbol{\mu}_C^{c0} \cdot \boldsymbol{R}_A \times \boldsymbol{\mu}_A^{01}\right\}$$

$$= \frac{-\varepsilon_a \varepsilon_c V^{1c}}{\hbar\left(\varepsilon_c^2 - \varepsilon_a^2\right)}\left\{\boldsymbol{\mu}_C^{0c} \times \boldsymbol{\mu}_A^{01} \cdot \boldsymbol{R}_{AC}\right\} \tag{7.52}$$

As with the degenerate coupled-oscillator equations, the application of eqn (7.52) is covered in Chapter 5.

For the particular case where one chromophore, **A**, non-degenerately couples to n-identical chromophores arranged around it so **A** lies on the z-axis of the n-chromophore system, then the coupled-oscillator *CD* follows simply by inserting a summation over n into eqn (7.52). Depending on the symmetry of **A** and the polarization of the **A** transition the equation can be further simplified as summarized in Table 7.1[6] and as was illustrated in Chapter 5 for cyclodextrin inclusion compounds.

Table 7.1 Coupled-oscillator CD expressions at energy ε_a for transitions of a guest molecule **A** at the centre of a host composed of n identical **C** chromophores arranged so the host has an n-fold rotational symmetry axis.

Host	Guest	CD expression
$n = 2$	non-degenerate	$R(x) = n\mu_{Ax}\mu_{Ax}H_{yzx}$
	$\mathbf{C}_{2v}$, $\mathbf{D}_{2h}$ symmetry	$R(y) = n\mu_{Ay}\mu_{Ay}H_{zxy}$
		$R(z) = n\mu_{Az}\mu_{Az}H_{xyz}$
	u/v degenerate	$R(u/v) = -\dfrac{n}{2}\left(\mu_{Au}\mu_{Au} + \mu_{Av}\mu_{Av}\right)H_{uvw}$
		$R(w) = n\mu_{Aw}\mu_{Aw}H_{uvw}$
$n > 2$	non-degenerate	$R(x) = -\dfrac{n}{2}\mu_{Ax}\mu_{Ax}H_{xyz}$
x/y-degenerate		$R(y) = -\dfrac{n}{2}\mu_{Ay}\mu_{Ay}H_{xyz}$
		$R(z) = n\mu_{Az}\mu_{Az}H_{xyz}$
	x/y-degenerate	$R(x/y) = -\dfrac{n}{2}\left(\mu_{Ax}\mu_{Ax} + \mu_{Ay}\mu_{Ay}\right)H_{xyz}$
		$R(z) = n\mu_{Az}\mu_{Az}H_{xyz}$
	x/z-degenerate	$R(x/y) = -\dfrac{n}{4}\left(\mu_{Ax}\mu_{Ax} + \mu_{Az}\mu_{Az}\right)H_{xyz}$
		$R(z) = \dfrac{n}{2}\mu_{Ay}\mu_{Ay}H_{xyz}$

Guest transition polarizations are indicated in parentheses and state labels have been omitted for clarity. Let

$$H_{uvw} = \frac{\varepsilon_a \varepsilon_c}{\hbar R_{AC}^2 (\varepsilon_c^2 - \varepsilon_a^2)}\left(\mu_{Cu}\hat{\boldsymbol{R}}_{ACv} - \mu_{Cv}\hat{\boldsymbol{R}}_{ACu}\right)\mu_{Cw}$$

For the general case a summation over **C** transitions should be taken.

7.7 Qualitative approach to exciton *CD*

The coupled-oscillator (or exciton) theory is an extremely useful tool for determining the absolute configuration of a chiral compound. It is in fact, apart from X-ray crystallographic determination, the most reliable non-empirical method for this task. For the practical purpose of assignment of *CD* signs to a given enantiomeric form, a full quantitative calculation is often unnecessary and a qualitative analysis along the lines given below will suffice. In addition, this qualitative approach will provide us with a way to understand the physical significance of the equations derived above and used in Chapter 5.

Both the case of two non-degenerate transitions and that of two degenerate transitions (the true exciton case) provide *CD* signals given by a scalar triple product:

$$\boldsymbol{\mu}^{a} \cdot \boldsymbol{R}_{AC} \times \boldsymbol{\mu}^{c} \tag{7.53}$$

where $\boldsymbol{\mu}^{a}$ and $\boldsymbol{\mu}^{c}$ are the transition moments of two chromophore moieties that are chirally arranged relative to one another in space and $\boldsymbol{R}_{AC}$ is the distance vector connecting them. We may here see an analogy with the *CD* as theoretically provided by the Rosenfeld equation:

$$\boldsymbol{\mu} \cdot \boldsymbol{m} \tag{7.54}$$

The second vector, $\boldsymbol{R}_{AC} \times \boldsymbol{\mu}^{c}$, in eqn (7.53) may be regarded as the magnetic moment that the electric moment $\boldsymbol{\mu}^{c}$ exerts on $\boldsymbol{\mu}^{a}$ by orbiting at a radius of R_{AC} around the $\boldsymbol{\mu}^{a}$ centre.

We shall focus on the case of degenerate transitions where we have first to consider the effects that the electric moments have on each other. From coupling between the transition moments of the two degenerate transitions, the following exciton transition moments result

$$\boldsymbol{\mu}^{+} = \frac{1}{\sqrt{2}}\left(\boldsymbol{\mu}^{a} + \boldsymbol{\mu}^{c}\right)$$

$$\boldsymbol{\mu}^{-} = \frac{1}{\sqrt{2}}\left(\boldsymbol{\mu}^{a} - \boldsymbol{\mu}^{c}\right) \tag{7.55}$$

where the + and − signs correspond, respectively, to in-phase and out-of-phase combinations of the transition moments. These new transition moments are obtained, when considering the picture of an assumed model for one of the enantiomers of the molecule, simply by moving the vector $\boldsymbol{\mu}^{c}$ to the tip of $\boldsymbol{\mu}^{a}$, forming a resultant vector with parallel or antiparallel orientations (see Appendix 1). We refer to this operation as **Step 1**.

Step 2 requires identifying the energies at which the resulting + and − exciton transitions occur. This is usually easily done by an electrostatic argument. Draw the vectorial combination $\boldsymbol{\mu}^{+}$, put a negative charge at the beginning and a positive charge at the tip of each of $\boldsymbol{\mu}^{a}$ and $\boldsymbol{\mu}^{c}$. Then judge whether this is an energetically favourable arrangement. If so, then $\boldsymbol{\mu}^{+}$ occurs at the lower energy, *i.e.* at the longer wavelength, than the degenerate monomeric transitions.

In **Step 3** we determine the sign of the *CD* for each of the exciton transitions of eqn (7.55) as follows. Look first at the resultant electric dipole transition moment $\boldsymbol{\mu}^{+}$ in the geometry of the proposed enantiomer. Now determine whether the electric moments $\boldsymbol{\mu}^{a}$ and $\boldsymbol{\mu}^{c}$ give rise to any magnetic moments that are parallel (in which case the *CD* will be positive) or antiparallel (negative *CD*) with respect to μ^+. This task is equivalent to determining whether the electric moments give rise to rotation of charge around the direction of μ^+ that is anticlockwise when viewed from the tip of μ^+ or clockwise. Agreement with the experimental sign pattern means the proposed enantiomer is correctly identified, disagreement means the opposite absolute configuration is in fact the correct one.

The examples considered in §5.7 may be analysed in this manner. There we focused on requiring a right-handed arrangement for μ^c, μ^a, and R_{AC} (CAR) for a positive signal. As Fig. 5.6 indicates this is equivalent to μ^c circling anticlockwise about μ^a when viewed from the tip of μ^a, so the two approaches are equivalent.

An analogous qualitative analysis may be pursued for non-degenerate coupled-oscillator *CD*. Instead of μ^+ and μ^- we use the unperturbed edtms in the two chromophores. The sign of the signal induced into the higher energy transition is the same as that from that deduced in **Step 3** for μ^+.

7.8 Magnetic dipole allowed transitions: the dynamic coupling model

In this section we shall see the origin of the equations used in Chapter 6 to account for the *CD* induced into mda transitions. *Ab initio* calculations[7] have been used to show that the mechanisms arising from this perturbation theory approach, first developed by Höhn and Weigang,[8] do indeed account for the *CD* in chiral adamantanones. However, anything other than a qualitative analysis of the electronic effects in planar zig-zag molecules (§6.4) are not feasible within a chromophoric approach since these arise from a break down of the chromophoric approach.

An mda transition of an achiral chromophore, **A**, must 'borrow' some electric character in order to have a *CD* signal as discussed in Chapter 6. As with coupled-oscillator *CD*, the required perturbation arises as a result of the electrostatic coupling of **A** with the rest of the molecule. The general case is more difficult to deal with for mda transitions than for eda transitions since the coupling due to the first term in the Taylor series expansion of V [eqn (7.38)] often vanishes. We shall therefore explicitly evaluate V^{1c} in the equations below as late as possible in the derivations.

We also again assume that *transition* moments give significantly stronger effects than *permanent* moments. This leads to the 'dynamic coupling' model for the *CD* of mda transitions. Historically much more effort has been put into what is known as the 'static coupling' or 'one electron' mechanism (since one electron on **A** is assumed to be the only one to move) where the molecular framework provides a static perturbation *via* permanent moments. This mechanism certainly contributes to the net observed *CD*, however, its contribution is almost always smaller than that due to the dynamic coupling mechanism. More details about the static coupling model may be found in references 9–11.

The formalism for dealing with mda transitions is very similar to the one we used for the non-degenerate coupled-oscillator *CD* in the preceding section, we just need to remember that here the $|0\rangle \to |1\rangle$ transition of **A** is mda and edf so $m_A^{01} \neq 0$ but $\mu_A^{10} = 0$. The perturbed wavefunctions $|i\rangle = |00\rangle$ and $|f\rangle = |10\rangle$ are given to first order in V by eqns (7.47) and (7.48), from which it follows that[9,10]

$$\mu^{10} = \langle 00|\tilde{\mu}|10\rangle = \langle 00|\hat{\mu}_A + \hat{\mu}_C|10\rangle$$

$$= \left(00|\hat{\mu}_A|10\right) - \frac{\left(00|V|1c\right)}{\varepsilon_c + \varepsilon_a}\left(1c|\hat{\mu}_C|10\right) - \frac{\left(0c|V|10\right)}{\varepsilon_c - \varepsilon_a}\left(00|\hat{\mu}_C|0c\right)$$

$$= -\frac{\left(00|V|1c\right)}{\varepsilon_a + \varepsilon_c}\mu_C^{0c} - \frac{\left(0c|V|10\right)}{\varepsilon_c - \varepsilon_a}\mu_C^{c0} \qquad (7.56)$$

$$= -\frac{2\varepsilon_c V^{1c}}{\varepsilon_c^2 - \varepsilon_a^2}\mu_C^{c0}$$

since the **A** and **C** wavefunctions are real. Similarly,

$$m^{01} = \langle 10|\tilde{m}|00\rangle = \langle 10|\hat{m}_A + \hat{m}_C|00\rangle$$

$$= \left(10|\hat{m}_A|00\right) - \frac{\left(1c|V|00\right)}{\varepsilon_a + \varepsilon_c}\left(10|\hat{m}_C|1c\right) - \frac{\left(10|V|0c\right)}{\varepsilon_c - \varepsilon_a}\left(0c|\hat{m}_C|00\right)$$

$$= m_A^{01} - \frac{\left(1c|V|00\right)}{\varepsilon_a + \varepsilon_c}m_C^{c0} - \frac{\left(10|V|0c\right)}{\varepsilon_c - \varepsilon_a}m_C^{0c} \qquad (7.57)$$

$$= m_A^{01} - \frac{2\varepsilon_a V^{1c}}{\varepsilon_c^2 - \varepsilon_a^2}m_C^{0c}$$

Substitution into the Rosenfeld equation gives the so-called dynamic coupling *CD* expressions to first order in V.

$$R = \mathrm{Im}\left\{\frac{-2\varepsilon_c V^{1c}}{\varepsilon_c^2 - \varepsilon_a^2}\mu_C^{c0} \cdot m_A^{01}\right\} \qquad (7.58)$$

In order to apply eqn (7.58) we need to have explicit forms for V^{1c} in terms of electric transition moments of **A** and **C**. It is now convenient to use a general form for the terms in the expansion of V using the Kronecker delta, δ_{ij}, which is zero unless $i = j$. We rewrite eqn (7.42), the dipole–dipole term in the expansion of V as

$$V^{1c}\left(\mu_A - \mu_C\right) = \frac{1}{R_{AC}^3}\left[\left(\mu_A^{10}\right)_i\left(\mu_C^{0c}\right)_j\right]\left[\delta_{ij} - 3\left(\hat{R}_{AC}\right)_i\left(\hat{R}_{AC}\right)_j\right] \qquad (7.59)$$

where a sum over all repeated indices is implied. When substituted into eqn (7.58), the dipole–dipole term of V leads to a *CD* expression containing a $\left(\mu_A^{01}\right)_i\left(m_A^{10}\right)_k$ term. This moment product vanishes except for chromophores of $\mathbf{C}_s$ symmetry and x- and y-polarized transitions of $\mathbf{C}_{2v}$ molecules. If a chromophore has $\mathbf{C}_s$ symmetry then the **A** moment product does not vanish; however, an mda transition in such a molecule is also eda so the *CD* will almost certainly be dominated by the coupled-oscillator *CD* of the edtm which is dependent on $\left(R_{AC}\right)^{-2}$ (see above). The one case where we cannot assume that a coupled-oscillator *CD* dominates was discussed in Chapter 6: mda d-d transitions even of low symmetry transition metal complexes have very small edtms due to the intrinsic symmetry of the d orbitals.

Thus the dynamic coupling mechanism first (in order of increasing chromophore symmetry) becomes important for transitions with z-polarized mdtms of $\mathbf{C}_{2v}$ chromophores such as the carbonyl which formed the subject matter of much of Chapter 6. The quadrupole–dipole term of V is the first non-vanishing one here:

$$V^{1c}(Q_A - \mu_C) = \frac{3}{2R_{AC}^4}\left[\left(Q_A^{10}\right)_{ij}\left(\mu_C^{0c}\right)_k\right]$$

$$\times\left[\begin{array}{l}\delta_{ij}\left(\hat{R}_{AC}\right)_k + \delta_{ik}\left(\hat{R}_{AC}\right)_j + \delta_{jk}\left(\hat{R}_{AC}\right)_i \\ -5\left(\hat{R}_{AC}\right)_i\left(\hat{R}_{AC}\right)_j\left(\hat{R}_{AC}\right)_k\end{array}\right] \tag{7.60}$$

where $\hat{Q}_A$ is the electric quadrupole moment operator of **A**. It is a second rank tensor whose components are

$$\left(\hat{Q}_A\right)_{ij} = e\sum_a\left(r_i r_j\right) \tag{7.61}$$

[*cf.* eqn (7.7) for notation]. The quadrupole–dipole *CD* strength is thus

$$R = -\frac{3}{R_{AC}^4}\,\text{Im}\left[\left(Q_A^{10}\right)_{ij}\left(m_A^{01}\right)_\ell\right]\left[\frac{\varepsilon_c\left(\mu_C^{c0}\right)_\ell\left(\mu_C^{0c}\right)_k}{\left(\varepsilon_c^2 - \varepsilon_a^2\right)}\right]$$

$$\times\left[\begin{array}{l}\delta_{ij}\left(\hat{R}_{AC}\right)_k + \delta_{ik}\left(\hat{R}_{AC}\right)_j + \delta_{jk}\left(\hat{R}_{AC}\right)_i \\ -5\left(\hat{R}_{AC}\right)_i\left(\hat{R}_{AC}\right)_j\left(\hat{R}_{AC}\right)_k\end{array}\right] \tag{7.62}$$

For transitions with z-polarized mdtms, such as the $n \to \pi^*$ transition of carbonyls, $\ell = z$ in eqn (7.62). Further, under $\mathbf{C}_{2v}$ symmetry unless $\{i, j\}$ equals $\{x, y\}$ or $\{y, x\}$ the **A**-moment product vanishes. Thus for $\mathbf{C}_{2v}$ **A** the *CD* for z-polarized mda transition is:

$$R(\mathbf{C}_{2v}, z) = -\frac{6}{R_{AC}^4}\,\text{Im}\left[\left(Q_A^{01}\right)_{xy}\left(m_A^{10}\right)_z\right]\left[\frac{\varepsilon_c\left(\mu_C^{0c}\right)_z}{\left(\varepsilon_c^2 - \varepsilon_a^2\right)}\right]$$

$$\times\left[\begin{array}{l}\left(\mu_C^{c0}\right)_x\left(\hat{R}_{AC}\right)_y + \left(\mu_C^{c0}\right)_y\left(\hat{R}_{AC}\right)_x \\ -5\left(\mu_C^{c0}\right)_k\left(\hat{R}_{AC}\right)_x\left(\hat{R}_{AC}\right)_y\left(\hat{R}_{AC}\right)_k\end{array}\right] \tag{7.63}$$

In §6.4 we reduced our consideration to the term in eqn (7.63) that has an isotropic polarizability component (the octant rule term). As noted in Chapter 6, there are occasions when the octant rule is not appropriate because other terms in eqn (7.63) (or even higher order terms in the V expansion) become dominant.

Eqn (7.63) also leads to the *CD* expressions for degenerately x,y-polarized transitions of molecules with $\mathbf{D}_{3d}$ **A**, since the **A** moment product of eqn (7.62) does not vanish for these polarizations under this symmetry. Take z to be the three-fold rotation axis and x to align with one of the two-fold rotation axes of **A**. By symmetry, the only non-vanishing **A** moment products are those in eqn (7.64).[9]

$$R(\mathbf{D}_{3d}, x/y) = -\,\mathrm{Im}\,\frac{3\varepsilon_c}{4R_{AC}^4\left(\varepsilon_c^2 - \varepsilon_a^2\right)}$$

$$\times\left\{\begin{aligned}
&\left[\left(Q_A^{10}\right)_{xx}\left(m_A^{01}\right)_x - \left(Q_A^{10}\right)_{yy}\left(m_A^{01}\right)_x - 2\left(Q_A^{10}\right)_{xy}\left(m_A^{01}\right)_y\right] \\
&\times\left[\begin{aligned}
&2\left(\mu_C^{c0}\right)_x\left(\mu_C^{0c}\right)_x\left(\hat{R}_{AC}\right)_x - 5\left(\mu_C^{c0}\right)_k\left(\mu_C^{0c}\right)_x\left(\hat{R}_{AC}\right)_x\left(\hat{R}_{AC}\right)_x\left(\hat{R}_{AC}\right)_k \\
&-4\left(\mu_C^{c0}\right)_x\left(\mu_C^{0c}\right)_y\left(\hat{R}_{AC}\right)_y + 5\left(\mu_C^{c0}\right)_k\left(\mu_C^{0c}\right)_x\left(\hat{R}_{AC}\right)_y\left(\hat{R}_{AC}\right)_y\left(\hat{R}_{AC}\right)_k \\
&-2\left(\mu_C^{c0}\right)_y\left(\mu_C^{0c}\right)_y\left(\hat{R}_{AC}\right)_x + 10\left(\mu_C^{c0}\right)_k\left(\mu_C^{0c}\right)_x\left(\hat{R}_{AC}\right)_x\left(\hat{R}_{AC}\right)_y\left(\hat{R}_{AC}\right)_k
\end{aligned}\right] \\
&-4\left[\left(Q_A^{10}\right)_{yz}\left(m_A^{01}\right)_x - \left(Q_A^{10}\right)_{xz}\left(m_A^{01}\right)_y\right] \\
&\times\left[\begin{aligned}
&\left(\mu_C^{c0}\right)_y\left(\mu_C^{0c}\right)_z\left(\hat{R}_{AC}\right)_x + 5\left(\mu_C^{c0}\right)_k\left(\mu_C^{0c}\right)_x\left(\hat{R}_{AC}\right)_y\left(\hat{R}_{AC}\right)_z\left(\hat{R}_{AC}\right)_k \\
&-\left(\mu_C^{c0}\right)_z\left(\mu_C^{0c}\right)_x\left(\hat{R}_{AC}\right)_y - 5\left(\mu_C^{c0}\right)_k\left(\mu_C^{0c}\right)_y\left(\hat{R}_{AC}\right)_x\left(\hat{R}_{AC}\right)_z\left(\hat{R}_{AC}\right)_k \\
&+\left(\mu_C^{c0}\right)_x\left(\mu_C^{0c}\right)_y\left(\hat{R}_{AC}\right)_z - \left(\mu_C^{c0}\right)_y\left(\mu_C^{0c}\right)_x\left(\hat{R}_{AC}\right)_z
\end{aligned}\right]
\end{aligned}\right\}$$

$$\tag{7.64}$$

The isotropic polarizability [eqn (6.6)] part of eqn (7.64) is

$$R(\mathbf{D}_{3d})_{x,y} = \frac{15\varepsilon_c\alpha_C(\varepsilon_a)}{4R_{AC}^4\left(\varepsilon_c^2 - \varepsilon_a^2\right)}$$

$$\times\left\{\begin{aligned}
&\mathrm{Im}\left[\left(Q_A^{10}\right)_{xx}\left(m_A^{01}\right)_x - \left(Q_A^{10}\right)_{yy}\left(m_A^{01}\right)_x - 2\left(Q_A^{10}\right)_{xy}\left(m_A^{01}\right)_y\right] \\
&\times\left(\hat{R}_{AC}\right)_x\left[\left(\hat{R}_{AC}\right)_x^2 - 3\left(\hat{R}_{AC}\right)_y^2\right]
\end{aligned}\right\}$$

$$\tag{7.65}$$

Eqn (7.65) was the one used in the discussion of *d-d* transitions of tris-chelate metal complexes in Chapter 6.

When **A** has higher than two-fold rotational symmetry, then the quadrupole–dipole term in the expansion of *V* is also zero for transitions polarized along the high symmetry rotation axes. Since the octupole term in the *V* expansion also vanishes under these circumstances, the appropriate term in the *V* expansion is the hexadecapole–dipole term which has an R_{AC}^6 factor in the denominator. The appropriate equation may be found in reference 9.

The above equations may be used as the basis for qualitative comparisons of magnitudes expected for the *CD* induced into different transitions, and if the isotropic polarizability terms, with their comparatively simple sector rules are dominant, then more detailed relationships between geometry and *CD* may be deduced. The degree of parametrization or calculation required to implement the full equation for, say, a $\mathbf{D}_{3d}$ molecule means that these equations are unlikely to be used.

7.9 Magnetic circular dichroism

In §7.6 and §7.7 we derived the equations describing the *CD* induced into an eda transition by its chiral molecular environment. Another way of inducing helical character into a linear electron displacement is to impose a magnetic field on the system. In this section we shall derive the equations used in §6.6 which describe the effect of a permanent magnetic field on an eda transition. The presence of the external magnetic field means that we cannot work with real wavefunctions in this section and it is therefore now important to be careful with the order of superscripts on the edtms as well as on the mdtms. We proceed by taking the real wavefunctions of the molecule in the absence of the magnetic field and then determine the effect of the magnetic field using perturbation theory and the Zeeman perturbation

$$V = -\hat{\boldsymbol{\epsilon}}_k \cdot \hat{\boldsymbol{m}} \mathrm{B}_k \tag{7.66}$$

where $\hat{\boldsymbol{\epsilon}}_k$ is the unit vector along the radiation propagation direction (which we take to be the X-axis), B_k is the magnitude of the external magnetic field along the k axis, and $\hat{\boldsymbol{m}}$ is the molecule's magnetic dipole moment operator. Maximal *MCD* signals will be generated when the direction of the external magnetic field is also along X, so we take

$$\mathbf{B}_{\text{external}} = \mathrm{B}(1,0,0) \tag{7.67}$$

The wavefunction for, say, the ground state of a molecule in the presence of the external magnetic field may then be written as a sum of the unperturbed ground state wavefunction plus contributions from unperturbed excited states. The coefficient describing how much any one excited state contributes to the perturbed state is, to first order in V, purely imaginary since $\hat{\boldsymbol{m}}$ is imaginary [eqn (7.66)]. The *MCD* then follows upon finding expressions for $|0\rangle$ and $|k\rangle$, $k = 1, 2, ...,$ the states of the molecule in the presence of the magnetic field. The magnetic field that is imposed during an *MCD* experiment is small compared with other effects, so we can use perturbation theory to write expressions for the wavefunctions. This separates into two cases: degenerate and non-degenerate.

MCD for a transition from a non-degenerate ground state to a degenerate excited state

If the transition we are probing is degenerate (as is the case for, for example, T_{1g} transitions in octahedral transition metal complexes) then degenerate perturbation theory is used to express the coupling between the degenerate excited states in the presence of the magnetic field. The effect of the magnetic field on the non-degenerate ground state is comparatively small so is ignored. We limit consideration to two excited states, $|1\rangle$ and $|2\rangle$, that are degenerate in the absence of the magnetic field. Three or more are easily included by replacing $|1\rangle$ by $|j\rangle$ and $|2\rangle$ by $|k\rangle$ and summing over j and k in the final equations.

Degenerate perturbation theory involves solving the secular determinant for the degenerate states $|1\rangle$ and $|2\rangle$, which are by assumption real and orthonormal. The perturbation is $\hat{m}_X \mathrm{B}$, so the secular determinant is:

The Jahn–Teller effect precludes the existence of degenerate ground states, so we focus on degenerate excited states. If the Jahn–Teller distortion is small, then it may be valid to consider the ground state as degenerate in some examples. The analysis is similar to that presented here.

Note: many treatments of *MCD* begin by assuming we already know the functions that occur in the presence of the field.

$$\begin{vmatrix} E - \varepsilon_1 & -(1|\hat{m}_X B|2) \\ -(2|\hat{m}_X B|1) & E - \varepsilon_2 \end{vmatrix} = 0 \tag{7.68}$$

where E is the energy of the perturbed states and $\varepsilon_1 = \varepsilon_2 = \varepsilon$ are the energies of the unperturbed states. So

$$(E - \varepsilon)^2 = B^2 (1|\hat{m}_X|2)(2|\hat{m}_X|1) \tag{7.69}$$

from which it follows that

$$E = \varepsilon \mp iB(2|\hat{m}_X|1) \tag{7.70}$$

If we write the excited states in the form:

$$|f\rangle = c_1|1\rangle + c_2|2\rangle \tag{7.71}$$

the secular equation relates the coefficients c_1 and c_2 as follows

$$\pm c_1 iB(2|\hat{m}_X|1) = c_2 B(2|\hat{m}_X|1)$$
$$\pm c_1 i = c_2 \tag{7.72}$$

so the two states that result from the coupling of the degenerate states $|1\rangle$ and $|2\rangle$ when the magnetic field is present are:

$$E = \varepsilon - iB(2|\hat{m}_X|1); \quad |f_+\rangle = \frac{|1\rangle + i|2\rangle}{\sqrt{2}}$$

$$= \varepsilon + B\,\mathrm{Im}(2|\hat{m}_X|1)$$

$$= \varepsilon + B\,\mathrm{Im}\,m_X^{12}$$

$$E = \varepsilon + iB(2|\hat{m}_X|1); \quad |f_-\rangle = \frac{|1\rangle - i|2\rangle}{\sqrt{2}} \tag{7.73}$$

$$= \varepsilon - B\,\mathrm{Im}(2|\hat{m}_X|1)$$

$$= \varepsilon - B\,\mathrm{Im}\,m_X^{12}$$

To evaluate the *MCD* using eqn (7.23) we require the edtms of the perturbed transitions from the ground to excited states:

$$\boldsymbol{\mu}^{0f_\pm} = \langle f_\pm|\hat{\mu}|0\rangle = \frac{\{(1|\mp i(2|\}\hat{\mu}|0)}{\sqrt{2}} = \frac{\boldsymbol{\mu}^{01} \mp i\boldsymbol{\mu}^{02}}{\sqrt{2}} \tag{7.74}$$

and the reverse transition is the complex conjugate of eqn (7.74). So from eqn (7.23) we write the *MCD* rotatory strengths for two coupled transitions as

$$MCD(|0\rangle \to |f_\pm\rangle) = 2ik\left(\mu_Y^{f_\pm 0}\mu_Z^{0f_\pm} - \mu_Z^{f_\pm 0}\mu_Y^{0f_\pm}\right)$$

$$= ik\left\{ \begin{array}{l} \left(\mu_Y^{10} \pm i\mu_Y^{20}\right)\left(\mu_Z^{01} \mp i\mu_Z^{02}\right) \\ -\left(\mu_Z^{10} \pm i\mu_Z^{20}\right)\left(\mu_Y^{01} \mp i\mu_Y^{02}\right) \end{array} \right\}$$

$|1\rangle$ and $|2\rangle$ are real so

$$\left\{(1|\hat{m}_X|2)\right\}^* = (2|\hat{m}_X|1)$$
$$= -(1|\hat{m}_X|2)$$

$$MCD(|0\rangle \rightarrow |f_{\pm}\rangle) = ik\left\{ \begin{matrix} -\mu_Y^{01}\mu_Z^{10} \pm i\mu_Y^{02}\mu_Z^{10} \mp i\mu_Y^{01}\mu_Z^{20} - \mu_Y^{02}\mu_Z^{20} \\ +\mu_Z^{01}\mu_Y^{10} \mp i\mu_Y^{02}\mu_Y^{10} \pm i\mu_Z^{01}\mu_Y^{20} + \mu_Z^{02}\mu_Y^{20} \end{matrix} \right\}$$

$$= \pm 2k\left\{\mu_Y^{01}\mu_Z^{20} - \mu_Y^{02}\mu_Z^{10}\right\}$$

$$(7.75)$$

The last line of eqn (7.75) follows because the unperturbed states are real and all the edtms on its right side are in terms of unperturbed states. Eqn (7.75) may alternatively be written

$$MCD(|0\rangle \rightarrow |f_{\pm}\rangle) = \pm 2k\left\{\mu_Y^{01}\mu_Z^{20} - \mu_Y^{02}\mu_Z^{10}\right\}$$

$$= \pm 2k\left[\left(\boldsymbol{\mu}^{01} \times \boldsymbol{\mu}^{20}\right)\cdot\hat{\mathbf{B}}\right] \qquad (7.76)$$

$$= \pm 2k\left(\boldsymbol{\mu}^{01} \times \boldsymbol{\mu}^{20}\right)_X$$

Thus we expect two *MCD* bands of equal magnitude and opposite sign occurring at the energies given by eqn (7.70). As with the degenerate coupled-oscillator *CD* (§5.4), the *MCD* for degenerate transitions appears to be independent of the perturbation, in this case the external magnetic field. However, that is not the case since the energy separation of the two components depends on the applied magnetic field, and were we to include a complete treatment of bandshapes as is usually done with *MCD* (see *e.g.* reference 12), then the *MCD* would be obviously dependent on the applied field.

To measure an *MCD* spectrum we therefore require two degenerate transitions with different polarizations for which the mdtm between them is in yet another direction. Further, neither edtms must be parallel to the applied magnetic field. The usual experimental situation is to be measuring the *MCD* of a solution of randomly oriented molecules. In this case we note that the preceding equations are all in terms of the laboratory fixed axis. We also note that both the energy and the final *MCD* equations are in terms of the X-component of axial vectors. Thus if the molecule is rotated by $180°$ about an axis in the Y-Z plane the *MCD* signal inverts; however, as the energy perturbation changes sign simultaneously, rotational averaging does not cause the *MCD* signal to vanish so we can measure *MCD* on solutions.

MCD **for non-degenerate transitions**

Again, we limit our explicit consideration to three states on the unperturbed molecule (a summation can be introduced to generalize the result if needed): $|0\rangle$, the ground state, and two excited states $|1\rangle$ and $|2\rangle$ at energies ε_1 and ε_2, respectively, above the ground state. We choose the same experimental geometry as for the degenerate case and use non-degenerate perturbation theory

$$|0\rangle = |0\rangle + \mathrm{B}\left\{\frac{(1|\hat{m}_X|0)}{\varepsilon_1}|1\rangle + \frac{(2|\hat{m}_X|0)}{\varepsilon_2}|2\rangle\right\}$$

$$|1\rangle = |1\rangle + \mathrm{B}\left\{\frac{(0|\hat{m}_X|1)}{-\varepsilon_1}|0\rangle + \frac{(2|\hat{m}_X|1)}{\varepsilon_2 - \varepsilon_1}|2\rangle\right\} \qquad (7.77)$$

Note: The directions of the edtms may be arbitrarily chosen, but they define the direction of the mdtm in the energy expression, so reversing one edtm inverts the signs of both the *MCD* equations and the energies so the same experimental spectrum is predicted.

$$|2\rangle = |2\rangle + \mathrm{B}\left\{ \frac{(0|\hat{m}_X|2)}{-\varepsilon_2}|0\rangle + \frac{(1|\hat{m}_X|2)}{\varepsilon_1 - \varepsilon_2}|1\rangle \right\}$$

If we ignore all permanent moments, the required transition moments to evaluate the *MCD* for the perturbed system to first order in B are then:

$$\boldsymbol{\mu}(|0\rangle \to |1\rangle) = (1|\hat{\boldsymbol{\mu}}|0) + \mathrm{B}\left\{ (2|\hat{\boldsymbol{\mu}}|0)\frac{(1|\hat{m}_X|2)}{\varepsilon_2 - \varepsilon_1} + (1|\hat{\boldsymbol{\mu}}|2)\frac{(2|\hat{m}_X|0)}{\varepsilon_2} \right\}$$

$$= \boldsymbol{\mu}^{01} + \mathrm{B}\left\{ \boldsymbol{\mu}^{02}\frac{m_X^{21}}{\varepsilon_2 - \varepsilon_1} + \boldsymbol{\mu}^{21}\frac{m_X^{02}}{\varepsilon_2} \right\}$$

$$\boldsymbol{\mu}(|0\rangle \to |2\rangle) = (2|\hat{\boldsymbol{\mu}}|0) + \mathrm{B}\left\{ (1|\hat{\boldsymbol{\mu}}|0)\frac{(2|\hat{m}_X|1)}{\varepsilon_1 - \varepsilon_2} + (2|\hat{\boldsymbol{\mu}}|1)\frac{(1|\hat{m}_X|0)}{\varepsilon_1} \right\}$$

$$= \boldsymbol{\mu}^{02} + \mathrm{B}\left\{ \boldsymbol{\mu}^{01}\frac{m_X^{12}}{\varepsilon_1 - \varepsilon_2} + \boldsymbol{\mu}^{12}\frac{m_X^{01}}{\varepsilon_1} \right\}$$

(7.78)

From eqns (7.23) and (7.78) it follows that the *MCD* is

$$MCD(|0\rangle \to |1\rangle)$$

$$= \mathrm{i}2k\mathrm{B}\left\{ \begin{array}{l} \mu_Y^{10}\mu_Z^{02}\dfrac{m_X^{21}}{\varepsilon_2 - \varepsilon_1} + \mu_Y^{10}\mu_Z^{21}\dfrac{m_X^{02}}{\varepsilon_2} + \mu_Y^{20}\mu_Z^{01}\dfrac{m_X^{12}}{\varepsilon_2 - \varepsilon_1} \\[2mm] + \mu_Y^{21}\mu_Z^{01}\dfrac{m_X^{20}}{\varepsilon_2} - \mu_Z^{10}\mu_Y^{02}\dfrac{m_X^{21}}{\varepsilon_2 - \varepsilon_1} - \mu_Z^{10}\mu_Y^{21}\dfrac{m_X^{02}}{\varepsilon_2} \\[2mm] - \mu_Z^{20}\mu_Y^{01}\dfrac{m_X^{12}}{\varepsilon_2 - \varepsilon_1} - \mu_Z^{12}\mu_Y^{01}\dfrac{m_X^{20}}{\varepsilon_2} \end{array} \right\}$$

(7.79)

$$= \mathrm{i}4k\mathrm{B}\left\{ -\left(\mu_Y^{10}\mu_Z^{02} - \mu_Z^{10}\mu_Y^{02}\right)\frac{m_X^{12}}{\varepsilon_2 - \varepsilon_1} - \left(\mu_Y^{10}\mu_Z^{21} - \mu_Z^{10}\mu_Y^{21}\right)\frac{m_X^{20}}{\varepsilon_2} \right\}$$

and

$$MCD(|0\rangle \to |2\rangle)$$

$$= \mathrm{i}2k\mathrm{B}\left\{ \begin{array}{l} \mu_Y^{20}\mu_Z^{01}\dfrac{m_X^{12}}{\varepsilon_1 - \varepsilon_2} + \mu_Y^{20}\mu_Z^{12}\dfrac{m_X^{01}}{\varepsilon_1} + \mu_Y^{10}\mu_Z^{02}\dfrac{m_X^{21}}{\varepsilon_1 - \varepsilon_2} \\[2mm] + \mu_Y^{21}\mu_Z^{02}\dfrac{m_X^{10}}{\varepsilon_1} - \mu_Z^{20}\mu_Y^{01}\dfrac{m_X^{12}}{\varepsilon_1 - \varepsilon_2} - \mu_Z^{20}\mu_Y^{12}\dfrac{m_X^{01}}{\varepsilon_1} \\[2mm] - \mu_Z^{10}\mu_Y^{02}\dfrac{m_X^{21}}{\varepsilon_1 - \varepsilon_2} - \mu_Z^{21}\mu_Y^{02}\dfrac{m_X^{10}}{\varepsilon_1} \end{array} \right\}$$

$$= \mathrm{i}4k\mathrm{B}\left\{ \begin{array}{l} -\left(\mu_Y^{20}\mu_Z^{01} - \mu_Z^{20}\mu_Y^{01}\right)\dfrac{m_X^{12}}{\varepsilon_2 - \varepsilon_1} \\[2mm] + \left(\mu_Y^{20}\mu_Z^{12} - \mu_Z^{20}\mu_Y^{12}\right)\dfrac{m_X^{01}}{\varepsilon_1} \end{array} \right\}$$

(7.80)

The final lines in the previous two equations follow because all the electric moments are real, and all the magnetic ones are purely imaginary.

For the usual situation where the sample is randomly oriented then the *MCD* of the $|0\rangle \rightarrow |1\rangle$ transition is

$$MCD(|0\rangle \rightarrow |1\rangle) = \frac{-\mathrm{i}4k\mathrm{B}}{3}\left\{ \frac{\boldsymbol{\mu}^{01} \times \boldsymbol{\mu}^{20} \cdot \boldsymbol{m}^{12}}{\varepsilon_2 - \varepsilon_1} + \frac{\boldsymbol{\mu}^{01} \times \boldsymbol{\mu}^{12} \cdot \boldsymbol{m}^{20}}{\varepsilon_2} \right\} \tag{7.81}$$

In most cases $(\varepsilon_2 - \varepsilon_1) \ll \varepsilon_2$, so the first term in eqn (7.76) is dominant. Thus

$$MCD(|0\rangle \rightarrow |1\rangle) = \frac{-\mathrm{i}4k\mathrm{B}}{3}\left\{ \frac{\boldsymbol{\mu}^{01} \times \boldsymbol{\mu}^{20} \cdot \boldsymbol{m}^{12}}{\varepsilon_2 - \varepsilon_1} \right\}$$

$$= \mathrm{Im}\, \frac{4k\mathrm{B}}{3}\left\{ \frac{\boldsymbol{\mu}^{01} \times \boldsymbol{\mu}^{20} \cdot \boldsymbol{m}^{12}}{\varepsilon_2 - \varepsilon_1} \right\} \tag{7.82}$$

The key feature of this equation is that we can only expect to observe *MCD* if a transition can couple with one that is not parallel to it. By permuting the labels '1' and '2' in the above equation and remembering $\boldsymbol{m}^{12} = -\boldsymbol{m}^{21}$ we find the *MCD* induced into the transition at energy ε_2. If the simple three-level system we have assumed is actually appropriate, then the *MCD* signals in the two transitions will be equal in magnitude and opposite in sign. Thus, we can use MCD to probe the relative polarizations of transitions, and also to see how many transitions there are in a given region of the spectrum.

The equations we have derived here correspond to what are commonly referred to as the 'B' terms.[12] They do not look exactly the same as those commonly quoted, however, since as discussed above we have made no attempt to consider the band shape.

References

(1) Craig, D. P.; Thirunamachandran, T. *Molecular quantum electrodynamics: An introduction to radiation-molecule interaction.* London: Academic Press, **1984**

(2) Barron, L. D. *Molecular light scattering and optical activity.* Cambridge: Cambridge University Press, **1982**

(3) Nordén, B. *Appl. Spectrosc. Rev.* **1978**, *14*, 157

(4) Nordén, B.; Kubista, M.; Kuruscev, T. *Q. Rev. Biophysics* **1992**, *25*, 51

(5) Atkins, P. W. *Molecular quantum mechanics.* Oxford: Oxford University Press, **1983**

(6) Schipper, P. E.; Rodger, A. *J. Amer. Chem. Soc.* **1983**, *105*, 4541

(7) Lightner, D. A.; Bouman, T. D.; Wijekoon, W. M. D.; Hansen, A. *J. Amer. Chem. Soc.* **1985**, *89*, 5805

(8) Höhn, E. G.; Weigang, O. E. *J. Chem. Phys.* **1968**, *48*, 1127

(9) Schipper, P. E.; Rodger, A. *Chem. Phys.* **1986**, *109*, 173

(10) Rodger, A.; Rodger, P. M. *J. Amer. Chem. Soc.* **1988**, *110*, 2361

(11) Richardson, F. S. *Chem. Rev.* **1979**, *79*, 18

(12) Michl, J.; Thulstrup, E. W. *Spectroscopy with polarized light;* New York: VCH, **1986**; Stephens, P. J. *Adv. Chem. Phys.* **1976**, *35*, 197

A1 Vectors

Vectors are used to describe any property that has both magnitude and direction. Pictorially they are represented by a straight arrow (whose length is proportional to the magnitude of the property) which points along the direction of the property. For exámple a car moving west with velocity 100 km/hour could be represented by an arrow 1 cm in length pointing to the left on a map oriented with north at the top. In this case the scale is 1 cm corresponding to 100 km/hour. A car moving at 200 km/hour in the same direction would be represented by a 2 cm arrow.

It is usually convenient to express vectors in terms of a right-handed $\{x, y, z\}$ Cartesian axis system. In this case the vector, v, is written $v = (X, Y, Z)$ where the point (X, Y, Z) is the Cartesian coordinate of the tip of the arrow head when the tail is sitting at the origin (Fig. A1.1a). The length of the vector is then

$$|v| = v = \sqrt{X^2 + Y^2 + Z^2} \tag{A1.1}$$

We define a unit vector, $\hat{v}$, to be the vector of unit length along the direction of v

$$\hat{v} = \frac{v}{v} \tag{A1.2}$$

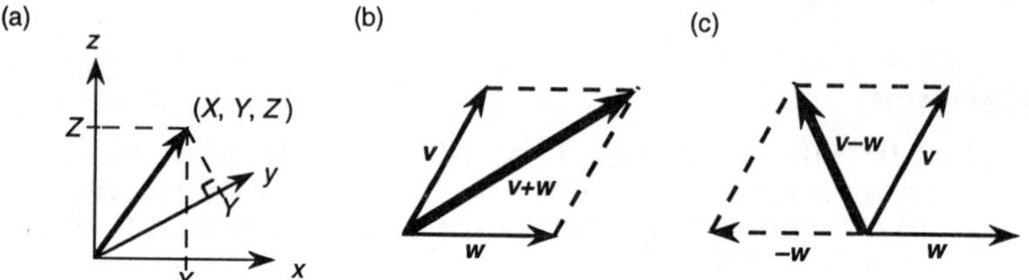

Fig. A1.1 (a) Cartesian representation of a vector. (b) Sum of two vectors. (c) Difference of two vectors.

Two vectors may be added together. Pictorially this is performed by determining the diagonal of the parallelogram as illustrated in Fig. A1.1b. Subtraction involves reversing the direction of the second vector (Fig. A1.1c). Mathematically for two vectors $v = \left(v_x, v_y, v_z\right)$ and $w = \left(w_x, w_y, w_z\right)$ this may be written

$$v + w = \left(v_x, v_y, v_z\right) + \left(w_x, w_y, w_z\right)$$
$$= \left(v_x + w_x, v_y + w_y, v_z + w_z\right) \tag{A1.3}$$

and

$$v - w = \left(v_x, v_y, v_z\right) - \left(w_x, w_y, w_z\right)$$

$$= \left(v_x - w_x, v_y - w_y, v_z - w_z\right) \tag{A1.4}$$

Three vector products are used in the text: the scalar product $v \cdot w$, the cross product $v \times w$, and the scalar triple product which is a combination of the other two: $u \cdot v \times w$. They are evaluated as follows.

$$v \cdot w = w \cdot v = \left(v_x, v_y, v_z\right) \cdot \left(w_x, w_y, w_z\right)$$

$$= v_x w_x + v_y w_y + v_z w_z \tag{A1.5}$$

$$v \times w = -w \times v = \left(v_x, v_y, v_z\right) \times \left(w_x, w_y, w_z\right)$$

$$= \begin{vmatrix} i & j & k \\ v_x & v_y & v_z \\ w_x & w_y & w_z \end{vmatrix} \tag{A1.6}$$

$$= \left(v_y w_z - v_z w_y, v_z w_x - v_x w_z, v_x w_y - v_y w_x\right)$$

where i, j, and k are, respectively, the unit vectors along the x-, y-, and z-axes.

$$u \cdot v \times w = v \times w \cdot u = u \times v . w$$

$$= -u \cdot w \times v = -u \times w . v$$

$$= \left(u_x, u_y, u_z\right) \cdot \left(v_x, v_y, v_z\right) \times \left(w_x, w_y, w_z\right)$$

$$= \begin{vmatrix} u_x & u_y & u_z \\ v_x & v_y & v_z \\ w_x & w_y & w_z \end{vmatrix} \tag{A1.7}$$

$$= u_x\left(v_y w_z - v_z w_y\right) - u_y\left(v_x w_z - v_z w_x\right) + u_z\left(v_x w_y - v_y w_x\right)$$

The scalar or dot product of two vectors is a measure of the projection of one vector onto the other. If they are parallel it is simply a product of their magnitudes, if they are antiparallel, then it has the same magnitude but is negative in sign, and if the vectors are perpendicular then the dot product is zero. The angle between the two vectors, τ is given by

$$\cos \tau = \frac{\mathbf{v} \cdot \mathbf{w}}{vw} \tag{A1.8}$$

The cross product of two vectors is a vector that is perpendicular to both of the original vectors, whose direction is such that the first vector, the second vector, and the new vector make a right-handed axis system (see marginal figure, p. 2). The magnitude of the new vector is the area of the parallelogram that the original two vectors define, so it is zero if those vectors are parallel.

The scalar triple product gives, as its name implies, a number which measures the volume of the parallelogram defined by the three vectors. So if the vectors are coplanar, the scalar triple product vanishes.

In the context of *CD* theory we often refer to axial vectors, which are as described above, and polar vectors. A polar vector is a vector that corresponds to a physical property, *e.g* angular momentum or magnetic dipole, and is defined as the cross product of two axial vectors. Because it is helpful to think of these vectors in terms of the original two we think of it as a circulation about the vector perpendicular to the plane defined by the original vectors with the direction defined to make a right-handed system as described above.

A2 Determination of equilibrium binding constants

The simplest measure of the binding strength between two molecules (such as a DNA binding ligand and DNA) is the equilibrium binding constant,

$$K = \frac{L_b}{L_f S_f} \tag{A2.1}$$

for the equilibrium

free ligand + empty binding site $\rightleftharpoons$ bound ligand

where L_b is the concentration of bound ligand, L_f is the concentration of free ligands, and S_f is the free site concentration. The total site concentration is

$$S_{tot} = \frac{C_M}{n} \tag{A2.2}$$

where C_M is the macromolecule concentration. For DNA we usually use the concentration of bases, in which case n is the number of bases in a binding site. For a protein, C_M is usually taken to be the concentration of protein molecules in which case

$$n' = \frac{1}{n} \tag{A2.3}$$

is the number of ligand binding sites on each protein molecule.

Sometimes for macromolecules one is reduced to evaluating the apparent binding constant at a given DNA and ligand concentration, since the binding strength and/or geometry is concentration dependent. However, concentration regimes either of a single binding mode or of constant proportions of a number of modes will often occur. K is indeed a constant in such uniform binding regimes, and the shapes of spectra such as that of the ligand induced CD or LD will also be constant. The four methods outlined below are particularly appropriate for use with LD and CD data.

The advantages of using spectroscopic data for determining K include the short timescale of the experiment, meaning that the system does not have to be stable for any great length of time, and our ability to probe a signal due only to the bound ligand. The starting point for equilibrium constant determination from spectroscopic data is usually the following equation:

$$L_b = \alpha \rho \tag{A2.4}$$

where ρ is the LD or CD signal at a chosen wavelength and α (which is a function of wavelength) is a constant over the range of binding ratios being considered. Wavelengths of maximum magnitude LD or CD signal will provide the most accurate data.

If binding reduces or increases S, the LD orientation parameter, in a non-linear fashion then that needs to be taken into account.

The simplest means of determining α is usually from the low binding ratio limit where all the ligands are assumed to be bound so L_f is assumed to be zero. If this is indeed the case then

$$\alpha = \frac{\rho}{L_{tot}} \qquad \text{(A2.5)}$$

where L_{tot} is the total ligand concentration. Alternatively, as indicated below in method II, the maximum LD or CD signal may be used to determine α if the binding site size, n, is known.

Eqn (A2.4) restricts us not only to regimes of uniform binding, but also to situations where ligand–ligand interactions do not affect the signal. This is unlikely to be a problem for LD, but if, for example, ligands stack externally on DNA their CD signal arises from ligand–ligand exciton coupling whose magnitude will depend non-linearly on the number of molecules in the stack.

I. Scatchard plot

The method most widely used in one form or another for determining K is the one developed by Scatchard.[1] Eqn (A2.1) is rearranged as follows:

$$\frac{r}{L_f} = \frac{KS_f}{C_M}$$
$$= \frac{K}{n} - rK \qquad \text{(A2.6)}$$

where

$$r = \frac{L_b}{C_M} \qquad \text{(A2.7)}$$

since $S_f = S_{tot} - L_b$. So, a plot of r/L_f versus r has slope K and y-intercept K/n. The x-intercept occurs where $r = n$. L_b, and hence L_f, may be determined directly from the LD (or CD) if α of eqn (A2.4) has been determined as discussed above. A Scatchard plot is illustrated below in Fig. A2.2c, where results from the intrinsic method (see below) are used to calculate the required r and L_f values.

II. Ligand number

This method was used by Nordén and Tjerneld[2] for LD, and it has since been independently used for normal absorption studies.[3] The ligand number, a concept that was originally introduced in the context of step-wise formation of metal–ligand complexes,[4] may be defined as:

$$v = \frac{L_b}{S_{tot}} = \frac{\rho}{\rho_{max}} \qquad \text{(A2.8)}$$

where ρ_{max} is the high ligand concentration limiting value for the LD or CD corresponding to the situation where every binding site is occupied so the maximum possible signal is being measured. In this case eqn (A2.1) is rewritten

$$K = \frac{L_b S_{tot}}{L_f S_{tot}(S_{tot} - L_b)} \qquad \text{(A2.9)}$$

which upon using the definition of v becomes

$$K = \frac{L_b}{L_f S_{tot}(1-v)} \tag{A2.10}$$

so

$$\frac{1}{(1-v)} = \frac{(L_{tot} - L_b)S_{tot}K}{L_b}$$

$$= \frac{L_{tot}}{v}K - \frac{C_M}{n}K \tag{A2.11}$$

Thus $\dfrac{1}{(1-v)}$ versus $\dfrac{L_{tot}}{v}$ has slope K, and y-intercept $-\dfrac{C_M}{n}K$.

III. Intrinsic method

It is not always possible to get data for either high or low binding ratio limits. In such cases eqn (A2.1) may be written

$$K = \frac{\alpha\rho}{(S_{tot} - \alpha\rho)(L_{tot} - \alpha\rho)} \tag{A2.12}$$

Rearranging eqn (A2.12) gives

$$L_{tot} = \frac{L_{tot}S_{tot}}{\alpha\rho} - S_{tot} + \alpha\rho - \frac{1}{K} \tag{A2.13}$$

so, for two different total ligand concentrations, L_{tot}^{j} and L_{tot}^{k}, but the same macromolecule concentration, *i.e.* $S_{tot}^{j} = S_{tot}^{k}$

$$\frac{L_{tot}^{k} - L_{tot}^{j}}{\rho^{k} - \rho^{j}} = \frac{S_{tot}}{\alpha}\left(\frac{\dfrac{L_{tot}^{k}}{\rho^{k}} - \dfrac{L_{tot}^{j}}{\rho^{j}}}{\rho^{k} - \rho^{j}}\right) + \alpha \tag{A2.14}$$

Thus a plot of

$$y = \frac{L_{tot}^{k} - L_{tot}^{j}}{\rho^{k} - \rho^{j}} \tag{A2.15}$$

versus

$$x = \left(\frac{\dfrac{L_{tot}^{k}}{\rho^{k}} - \dfrac{L_{tot}^{j}}{\rho^{j}}}{\rho^{k} - \rho^{j}}\right) \tag{A2.16}$$

(where any pair of data points considered have the same C_M) should be a straight line with slope $C_M(n\alpha)^{-1}$ and intercept α. The concentration of bound molecules in any sample may then be determined from eqn (A2.4) and the effective binding site size from eqn (A2.2). The equilibrium binding constant, K, follows directly from eqn (A2.12). Alternatively, using these accurate values of n and α a Scatchard plot (Method I) may be used to determine the best value of K using all the data points. The intrinsic method[5] is illustrated for some *CD* data in Fig. A2.1 and for some *LD* data from which a Scatchard plot is then used to determine K in Fig. A2.2.

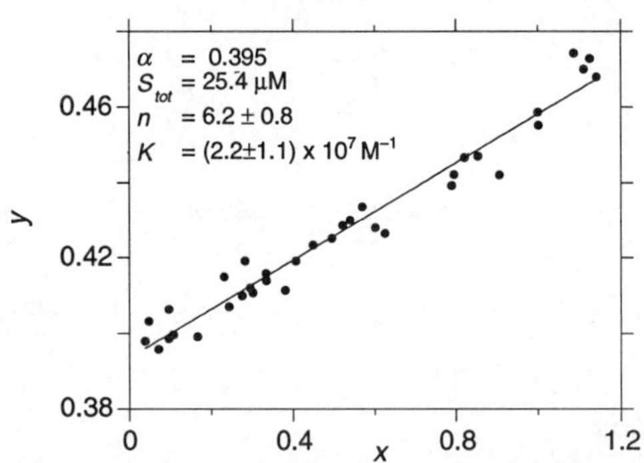

Fig. A2.1 Graph illustrating application of the intrinsic method for determining α and n using the data given in Fig. 2.12a. K has been determined using an average value from repeated application of eqn (A2.12). x and y are as defined in eqns (A2.15) and (A2.16).

It is sometimes convenient to perform experiments with constant ligand and varying macromolecule concentration. In this case C_M, and hence S_{tot}, are the variables and L_{tot} is fixed. Rather than eqn (A2.14) we then use:

$$\frac{S_{tot}^k - S_{tot}^j}{\rho^k - \rho^j} = \frac{L_{tot}}{\alpha}\left(\frac{\dfrac{S_{tot}^k}{\rho^k} - \dfrac{S_{tot}^j}{\rho^j}}{\rho^k - \rho^j}\right) + \alpha \tag{A2.17}$$

or equivalently

$$\frac{C_M^k - C_M^j}{\rho^k - \rho^j} = \frac{L_{tot}}{\alpha}\left(\frac{\dfrac{C_M^k}{\rho^k} - \dfrac{C_M^j}{\rho^j}}{\rho^k - \rho^j}\right) + n\alpha \tag{A2.18}$$

So the plot

$$y = \frac{C_M^k - C_M^j}{\rho^k - \rho^j} \tag{A2.19}$$

versus

$$x = \left(\frac{\dfrac{C_M^k}{\rho^k} - \dfrac{C_M^j}{\rho^j}}{\rho^k - \rho^j}\right) \tag{A2.20}$$

has slope $\dfrac{L_{tot}}{\alpha}$ and y-intercept $n\alpha$.

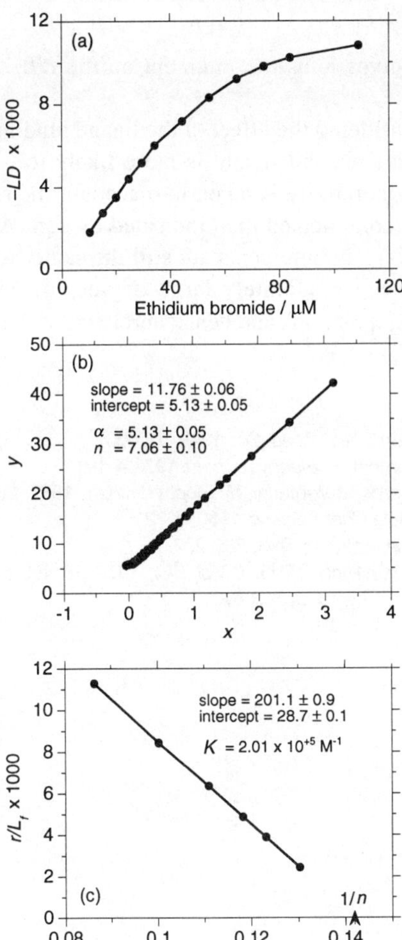

Fig. A2.2 Intrinsic method determination of α and n for the binding of ethidium bromide to calf thymus DNA (426 μM in base, in 0.2 M NaCl) and subsequent Scatchard plot. *LD* data from reference 2. (a) Experimental data, (b) intrinsic method plot, x and y as defined in eqns (A2.15) and (A2.16), and (c) Scatchard plot using data from final six points in (a) where the relationship between *LD* and total ligand concentration is not a straight line through the origin. K agrees with the best value from reference 2. Note all concentrations are plotted in units of μM, all other units are consistent between figures, and in (b) we use only the magnitude of the *LD*, so α is positive rather than negative.

IV. McGhee and von Hippel method

McGhee and von Hippel[6] analysed the binding of a ligand to a macromolecule, such as DNA, that could be represented as a lattice. Their concern was first to account for the fact that a binding site might span more than one lattice point, in which case defining $S = C_M/n$ is too simplistic except at low binding ratios (where this approach gives the same answer as the above methods). The revised form of eqn (A2.6) is then

$$\frac{r}{L_f} = K\left(1 - \frac{r}{n}\right)\left[\frac{n - r}{n - (1 - n)r}\right]^{\frac{1}{n} - 1} \qquad\qquad (A2.21)$$

so r/L_f versus r gives K as the intercept on the r/L_f axis and n on the r axis.

Further, they considered the effect of the ligand binding being cooperative, by which we mean a second ligand is more likely to bind next to the first than not. When cooperativity is taken into account the relationship between r/L_f and r is more complicated than indicated by eqn (A2.19) as outlined in reference 6, however, the intercepts are still the same. As noted by McGhee and von Hippel, even moderately large degrees of cooperativity lead to extremely steep extrapolations and hence unreliable values for K and n.

References

(1)   Scatchard, G. *Ann. N.Y. Acad. Sci.* **1949**, *51*, 660

(2)   Nordén, B.; Tjerneld, F. *Biophys. Chem.* **1976**,*4*, 191

(3)   Diebler, H.; Secco, F.; Venturini, M. *Biophys. Chem.* **1987**, *26*, 193

(4)   Fronaeus, S. *Acta Chem. Scand.* **1950**, *4*, 72

(5)   Rodger, A. *Meth. Enzymol.* **993**; *226*, 232

(6)   McGhee, J.D.; von Hippel, P.H. *J. Mol. Biol.*, **1974**, *86*, 469 and correction to key equation: *J. Mol. Biol.* **1976**, *103*, 679

A3 Momentum–dipole equivalence

One of the key equations [eqn (7.34)] used in deriving the coupled-oscillator
CD equations is the equivalence of the expectation value between two states,
$|j)$ and $|k)$ of the momentum operator for a chromophore, say **C**, and the
electric dipole transition moment from $|k)$ to $|j)$:

$$\left(k|\hat{\boldsymbol{p}}_{\mathrm{C}}|j\right) = \frac{im_e}{e\hbar}\left(\varepsilon_k - \varepsilon_j\right)\left(k|\hat{\boldsymbol{\mu}}_{\mathrm{C}}|j\right)$$

$$= \frac{im_e}{e\hbar}\left(\varepsilon_k - \varepsilon_j\right)\boldsymbol{\mu}_{\mathrm{C}}^{jk} \tag{A3.1}$$

Eqn (A3.1) follows from first noting that

$$\hat{\boldsymbol{p}} = -i\hbar\left(\frac{\partial}{\partial x}, \frac{\partial}{\partial y}, \frac{\partial}{\partial z}\right) = -i\hbar\,\boldsymbol{\nabla} \tag{A3.2}$$

Using the definition of $\boldsymbol{\mu}$ [eqn (7.6)], this means what we are required to
show is that

$$\left(k|\frac{\partial}{\partial x}|j\right) = -\frac{m_e}{\hbar^2}\left(\varepsilon_k - \varepsilon_j\right)\left(k|x|j\right) \tag{A3.3}$$

and similarly for y and z where the coordinates belong to chromophore **C**.
The derivation given here follows that of reference 1.

$|j)$ and $|k)$ satisfy the equations:

$$\frac{\partial^2}{\partial x^2}|k^*) + \frac{2m_e}{\hbar^2}\left[\varepsilon_k - V(x)\right]|k^*) = 0 \tag{A3.4}$$

$$\frac{\partial^2}{\partial x^2}|j) + \frac{2m_e}{\hbar^2}\left[\varepsilon_j - V(x)\right]|j) = 0 \tag{A3.5}$$

If we multiply eqn (A3.4) by $\left(j^*|x\right.$ and eqn (A3.5) by $\left(k|x\right.$ and then
subtract, we obtain

$$\left\{\left(\left(j|x\frac{\partial^2}{\partial x^2}|k\right)\right)^* - \left(k|x\frac{\partial^2}{\partial x^2}|j\right)\right\} = +\frac{2m_e}{\hbar^2}\left(\varepsilon_j - \varepsilon_k\right)\left\{\left(k|x|j\right)\right\} \tag{A3.6}$$

The first two terms may now be integrated by parts; since the wavefunctions
vanish at infinity, we have

$$\left\{\left(\frac{\partial}{\partial x}\left[\left(j|x\right]\frac{\partial}{\partial x}|k\right)\right)^* - \left(\frac{\partial}{\partial x}\left[\left(k|x\right]\frac{\partial}{\partial x}|j\right)\right)\right\} = +\frac{2m_e}{\hbar^2}\left(\varepsilon_k - \varepsilon_j\right)\left\{\left(k|x|j\right)\right\} \tag{A3.7}$$

or

$$\left\{\left(j\left|\frac{\partial}{\partial x}\right|k\right)^* - \left(k\left|\frac{\partial}{\partial x}\right|j\right)\right\} = +\frac{2m_e}{\hbar^2}\left(\varepsilon_k - \varepsilon_j\right)\left(k|x|j\right) \tag{A3.8}$$

Integration by parts can be used to show that the two terms on the left-hand side of eqn (A3.8) are equivalent. So

$$\left(k\left|\frac{\partial}{\partial x}\right|j\right) = -\frac{m_e}{\hbar^2}\left(\varepsilon_k - \varepsilon_j\right)\left\{\left(k|x|j\right)\right\} \tag{A3.9}$$

which is eqn (A3.3).

Reference

(1)   Eyring, H.; Walter, J.; Kimball, G. E. *Quantum Chemistry*. New York: John Wiley and Sons, **1944**

A4 Definitions and units

Linearly polarized light

We have defined the direction of propagation to be X. Light linearly polarized along the Y or Z axis has its electric field parallel to the unit vectors

$$\hat{\mathbf{e}}_Y = (0,1,0) \tag{A4.1}$$

or

$$\hat{\mathbf{e}}_Z = (0,0,1) \tag{A4.2}$$

Circularly polarized light

Right circularly polarized light describes a right-handed helix in space and rotates clockwise in time when viewed at a given point along X *towards* the source. Its electric field is parallel to

$$\hat{\mathbf{e}}_r = \frac{1}{\sqrt{2}}\left\{(0,1,-i)\exp\left(\frac{2\pi i X}{\lambda} - i\omega t\right)\right\} \tag{A4.3}$$

using the notation of eqn (7.3). Similarly, left circularly polarized light describes a left-handed helix in space and rotates anticlockwise in time when viewed towards the light source

$$\hat{\mathbf{e}}_\ell = \frac{1}{\sqrt{2}}\left\{(0,1,i)\exp\left(\frac{2\pi i X}{\lambda} - i\omega t\right)\right\} \tag{A4.4}$$

Absorbance

The absorbance of a sample is defined to be the logarithm (usually to base 10) of the ratio of the incident and transmitted radiation

$$A = A_{10} = \log_{10}\left(\frac{I_o}{I}\right) \tag{A4.5}$$

Circular dichroism

CD in absorbance units is

$$CD = A_\ell - A_r \tag{A4.6}$$

Alternatively in molar units it is

$$CD = \varepsilon_\ell - \varepsilon_r \tag{A4.7}$$

where ε_r is the extinction coefficient for the absorption of right circularly polarized light. The units used for extinction coefficients are almost always absorbance units ($\text{mol}^{-1}\ \text{dm}^3\ \text{cm}^{-1}$) which gives values that are ten times greater than those in SI units. We shall retain this set of units for extinction coefficients. The Beer–Lambert law relates absorbance units to the extinction coefficient:

$$(A_\ell - A_r) = (\varepsilon_\ell - \varepsilon_r)C\ell \tag{A4.8}$$

where C is in units of mol dm^{-3} and ℓ is the pathlength in cm.

Ellipticity (old measure of *CD*)

The ellipticity, θ, is obtained from the ratio of the minor and major axes of the ellipse traced out by the electric field vector of the elliptically polarized light when it emerges from the circularly dichroic sample onto which linearly polarized light was incident.

$$\tan\theta = \tanh\left(\frac{\pi\ell}{\lambda}\left(n'_\ell - n'_r\right)\right) \tag{A4.9}$$

where n'_ℓ is the absorption index for left circularly polarized light. The wavelength, λ, must be in the same units as ℓ. For small ellipticities

$$\theta\,(\text{radians}) \approx \frac{\pi\ell}{\lambda}\left(n'_\ell - n'_r\right)$$
$$= \frac{2.303\ell C}{4}\left(\varepsilon_\ell - \varepsilon_r\right) \tag{A4.10}$$

so

$$\theta\,(\text{degrees}) = 32.98\,C\ell\left(\varepsilon_\ell - \varepsilon_r\right)$$
$$= 32.98\left(A_\ell - A_r\right) \tag{A4.11}$$

Specific   ellipticity

Another old measure of *CD* is the specific ellipticity

$$[\theta] = \frac{\theta}{C_g d} \tag{A4.12}$$

where C_g is sample concentration in g cm^{-3} and d is pathlength in dm.

Molar   ellipticity

The related molar ellipticity is

$$M_\theta = \frac{[\theta]M}{100} = \frac{100\theta}{C\ell} \tag{A4.13}$$

where M is molar mass in g, C is in moles dm^{-3}, and ℓ is in cm. Thus

$$M_\theta = 3298\left(\varepsilon_\ell - \varepsilon_r\right)$$
$$= 3298\Delta\varepsilon \tag{A4.14}$$

where ε_r is the extinction coefficient for the absorption of right circularly polarized light.

Optical activity (optical rotation, *OR*)

Optical rotation is the difference in refractive indices of left and right circularly polarized light upon passing through the medium:

$$OR = (n_\ell - n_r)\frac{\pi d}{\lambda}\text{(radians)} \qquad (A4.15)$$

where $(n_\ell - n_r)$ is called the *circular birefringence* and λ and d have the same units (usually dm are used).

OR is usually measured by determining the rotation of linearly polarized light upon passing through the solution. If the linearly polarized light is rotated clockwise when viewed into the light source the *OR* is called a positive or right (dextro) *OR*.

Optical rotation as a function of λ is called *optical rotatory dispersion* (*ORD*). *ORD* makes an S-shaped curve centred at the *CD* maximum (so-called anomalous optical rotatory dispersion). The *ORD* is also non-zero away from an absorption band, hence α_D values (the *ORD* at the sodium D line) may generally be used to characterize the enantiomeric excess of a solution. For a positive *CD* band the long-wavelength side of the *ORD* curve shows a larger (positive) *ORD* contribution, for a negative *CD* band a smaller (negative) *ORD* contribution.

Dipole moments

Electric dipole moment

The electric dipole moment as defined in eqn (7.7) is

$$\boldsymbol{\mu} = \sum_i \left(q_i \boldsymbol{r}_i \right) \qquad (A4.16)$$

where q_i is the charge of the ith particle located at the end of vector $\boldsymbol{r}_i$. The charge on an electron is $e = -1.602 \times 10^{-19}$ C in SI units. The units for the electric dipole moment are: C m or Debye. These are related by:

$$1\ \text{Debye (D)} = 3.336 \times 10^{-30}\ \text{C m} \qquad (A4.17)$$

Dipole moments of small molecules are typically about 1 D.

Magnetic dipole moment

In the absence of a magnetic field, the magnetic dipole moment of a collection of charges is:

$$m = \sum_i \left\{ \frac{q_i}{2m_i} \boldsymbol{r}_i \times \boldsymbol{p}_i \right\} \qquad (A4.18)$$

where m_i is the mass of the particle. If the particles are electrons, the factor $e/2m_e$ is often replaced by $-\mu_B / \hbar$ where

$$\mu_B = \frac{|e|\hbar}{2m_e} \qquad (A4.19)$$

$$= 9.273 \times 10^{-24}\ \text{m}^2\ \text{s}^{-1}\ \text{C}$$

is the Bohr magneton. The Bohr magneton is often used as the unit for magnetic moments.

Normal absorption, dipole strength, and oscillator strength

The integrated absorption coefficient over an entire absorption band is

$$\varepsilon_I = \int \varepsilon(v)\,dv \qquad (A4.20)$$

It may be related to the dipole strength, D^{01}, and the dimensionless oscillator strength, f^{01}, for a transition from state $|0\rangle$ to state $|1\rangle$ of a solution of randomly oriented molecules as follows.

$$D^{01} = \left|\boldsymbol{\mu}^{01}\right|^2 \qquad (A4.21)$$

Since experimental reality is that no transition occurs at a precisely defined frequency, but over a range, we write the dipole strength per molecule as

$$D^{01} = \frac{3\ln 10\, c\varepsilon_0 \hbar}{\pi N_A} \int \frac{\varepsilon(v)}{10v}\,dv$$
$$= 1.022 \times 10^{-61} \int \frac{\varepsilon(v)}{v}\,dv \qquad (A4.22)$$

which has units $C^2\ m^2$ if $\varepsilon_0 = 8.854 \times 10^{-12}\ kg^{-1}\ m^{-3}\ s^2\ C^2$ is the permittivity of a vacuum, N_A is Avagadro's number, and ε is left in its usual non-SI units (hence the factor of 10 in the denominator). The dimensionless oscillator strength is then

$$f^{01} = \frac{4\pi m_e}{3\hbar e^2} v\left|\boldsymbol{\mu}^{01}\right|^2$$
$$= \frac{4 m_e c\varepsilon_0 \ln 10}{N_A e^2} \int \frac{\varepsilon(v)}{10}\,dv \qquad (A4.23)$$

Rotatory (rotational or *CD*) strength

The Rosenfeld equation [eqn (7.22)] for the *CD* strength of the transition from state $|0\rangle$ to state $|1\rangle$ of a solution of randomly oriented molecules is

$$R^{01} = \mathrm{Im}\langle 0|\tilde{\boldsymbol{\mu}}|1\rangle \cdot \langle 1|\tilde{\boldsymbol{m}}|0\rangle \qquad (A4.24)$$

Analogously to the dipole strength, the *CD* strength is

$$R^{01} = \frac{3\ln 10\, c^2 \varepsilon_0 \hbar}{4\pi N_A} \int \frac{\varepsilon_\ell(v) - \varepsilon_r(v)}{10v}\,dv$$
$$= 7.659 \times 10^{-54} \int \frac{\varepsilon_\ell(v) - \varepsilon_r(v)}{v}\,dv \qquad (A4.25)$$

which has units of $C^2\ m^3 s^{-1}$ if ε is in its usual units.

The relative intensities of the *CD* and dipole strengths gives a convenient estimate of how easy it will be to measure a *CD* spectrum. The ratio is known as the dissymmetry factor. When SI units are used it is

$$g^{01} = \frac{4R^{01}}{cD^{01}} \qquad (A4.26)$$

Interaction energies

Coulomb's Law for the interaction energy between two charged particles separated by a distance R_{12} in SI units (J) is

$$V = \frac{q_1 q_2 \mu_0 c^2}{4 R_{12}}$$

$$= \frac{q_1 q_2}{4\pi\varepsilon_0 R_{12}} \tag{A4.27}$$

where $\mu_0 = 4 \times 10^{-7}$ kg m C^{-2} is the permeability of free space, R_{12} is in m, the charge is in C, and $c = 2.9979 \times 10^8$ m s^{-1}. To express the dipole–dipole interaction energy in SI units we write [*cf.* eqn (7.38)]

$$V = \frac{\boldsymbol{\mu}_1 \cdot \boldsymbol{\mu}_2 - 3\hat{\mathbf{R}}_{12} \cdot \boldsymbol{\mu}_1 \boldsymbol{\mu}_2 \cdot \hat{\mathbf{R}}_{12}}{4\pi\varepsilon_0 R_{12}^3} \tag{A4.28}$$

where R_{12} is in m, $\boldsymbol{\mu}_1$ and $\boldsymbol{\mu}_2$ in C m, so V is in units of J.

For example, two parallel dipoles of strength 1 D that are aligned (so perpendicular to the line connecting them) separated by a distance of 1 nm have a dipole–dipole interaction energy of

$$V = \frac{(3.34 \times 10^{-30})^2}{4\pi \times 8.85 \times 10^{-12} \times 10^{-27}} \tag{A4.29}$$

$$= 1.00 \times 10^{-22} \text{ J}$$

General references

- Craig, D. P.; Thirunamachandran, T. *Molecular quantum electrodynamics: An introduction to radiation-molecule interaction*. London: Academic Press, **1984**
- Barron, L. D. *Molecular light scattering and optical activity*. Cambridge: Cambridge University Press, **1982**
- Hollas, J. M. *Modern spectrocopy*. Chichester: John Wiley and Sons, **1987**
- Atkins, P. W. *Physical Chemistry*, 4th ed. Oxford: Oxford University Press **1991**

Index